NEXT-GENERATION BATTERIES WITH SULFUR CATHODES

NEXT-GENERATION BATTERIES WITH SULFUR CATHODES

KRZYSZTOF JAN SICZEK

ELSEVIER

ACADEMIC PRESS
An imprint of Elsevier

Academic Press is an imprint of Elsevier
125 London Wall, London EC2Y 5AS, United Kingdom
525 B Street, Suite 1650, San Diego, CA 92101, United States
50 Hampshire Street, 5th Floor, Cambridge, MA 02139, United States
The Boulevard, Langford Lane, Kidlington, Oxford OX5 1GB, United Kingdom

Library of Congress Cataloging-in-Publication Data
A catalog record for this book is available from the Library of Congress

British Library Cataloguing-in-Publication Data
A catalogue record for this book is available from the British Library

ISBN 978-0-12-816392-4

Publisher: Joe Hayton
Acquisition Editor: Lisa Reading
Editorial Project Manager: Andrae Akeh
Production Project Manager: Anitha Sivaraj
Cover Designer: Greg Harris

Typeset by SPi Global, India

Cover credit: Modified from Research Area, Advanced Battery Materials Lab.

CONTENTS

INTRODUCTION

The current development of electric and hybrid cars is most often associated with the use of various lithium-ion batteries for the collection of electricity. These batteries are composed of lithium-ion cells that have many advantages:

- high specific energy and load capacity, especially compared to acid batteries,
- long cycles and extended shelf life, with virtually no maintenance,
- high capacity, low internal resistance, good coulombic efficiency,
- Simple charging algorithm and relatively short charging time,
- Low self-discharge compared to NiCd and NiMH batteries.

However, these batteries also have a number of limitations:

- require thermal protection circuit,
- subject to degradation at high temperatures and at high voltages,
- characterized by slow charging at freezing ($<0°C$),
- have restrictions on transport regulations for the transport of larger quantities of such batteries.

In addition to the lithium-ion cells, new types of cells with high energy density have emerged recently. They use different types of electrodes and electrolytes. One of the interesting groups receiving more and more attention lately are batteries using different types of sulfur cathodes, the largest subgroup in this category being lithium-sulfur batteries.

In the first chapter of this monograph, basic concepts are presented.

CHAPTER ONE

Basic Concepts

According to [1], the battery is a device that converts chemical energy into electricity and vice versa. This summary is an introduction to the terminology used to describe, classify, and compare batteries. It defines the variables used to characterize battery operating conditions and describes the manufacturer's specifications used to characterize the nominal and maximum battery characteristics.

1.1 GENERAL CONCEPTS OF BATTERIES

The following list explains some general battery concepts.

- Cells, modules, and packages are high-voltage rechargeable batteries, consisting of individual modules and cells organized (connected) in series and parallel. The cell is the smallest packed form of the battery and usually provides a voltage range of 1–6 V. The module consists of several cells generally connected either in parallel or in series. The battery pack is assembled by combining the modules together, again either in series or in parallel.
- Battery classification: Not all batteries are equivalent, even batteries based on the same chemistry. The main compromise in battery development is between power and energy: rechargeable batteries can be high power or high energy, but not at the same time. Manufacturers often classify batteries with these categories. Another frequently used classification is high durability, which means that the chemistry has been modified to provide more durability at the expense of power and energy.
- C- and E-rates: When describing a battery, the discharge current is often expressed numerically as a multiple of size C in order to normalize the battery capacity, which often varies greatly between batteries. A multiple of the size C is the measure of the current at which the battery is discharged, referring to the maximum capacity. The 1C-rate means that the discharge current will discharge the fully charged battery within 1 h. For a 100 Ah battery, it corresponds to a discharge current of 100 A. A 5C-rate for this battery corresponds to a current of 500 A, and a multiple

Next-Generation Batteries With Sulfur Cathodes
https://doi.org/10.1016/B978-0-12-816392-4.00001-3

of C/2-rate corresponds to a current of 50 A. Likewise, the value (multiple) of size E describes the discharge power. 1E-rate corresponds to the discharge power to discharge a fully charged battery within 1 h.
- Secondary and primary cells: Primary cells after discharge cannot be recharged. Secondary cells, after recharging, may be recharged (repeatedly).

1.2 DIMENSIONS CHARACTERIZING THE CONDITION OF THE BATTERY

This section describes some of the variables used to describe the current battery condition.
- State of charge (SOC) (%): Expresses the current battery capacity as a percentage of the maximum capacity. SOC is generally calculated using time-based integration to determine the change in battery capacity as a function of time.
- Depth of discharge (DOD) (%): The percentage of battery capacity to which the battery is discharged, expressed as a percentage of the maximum capacity. Discharging to at least 80% of the discharge depth is considered to be a deep discharge state.
- State of health (SOH) (%): Defines the current battery capacity relative to the maximum value in the specification: 100% SOH means that the battery has a capacity equal to that given by the manufacturer, and 50% means that the capacity has dropped to half the specified value.
- Terminal voltage (V): Voltage between battery terminals with applied load (i.e., in the form of a resistor). Voltage on terminals depends on SOC and discharge/charge current.
- Open circuit voltage (V): Voltage between battery terminals without applied load (e.g., in the form of a resistor). The voltage of the open circuit depends on the state of the battery charge, increasing with the state of the charge.
- Internal resistance: Resistance within the battery, generally different for charging and discharging, and also depends on the state of the battery. As the internal resistance increases, the efficiency of the battery decreases, and as the thermal stability is reduced even more, the more charged energy is converted into heat.

1.3 BATTERY SPECIFICATIONS

This section explains the quantities that can be seen on the technical specifications used to describe batteries.

- Nominal voltage (V): The specified or reference battery voltage, sometimes referred to as "normal" battery voltage.
- Cut-off voltage: The minimum permissible voltage. This is a voltage that generally defines the "empty" state of the battery, reached in the discharge process, upon which the battery becomes fully discharged.
- Capacity or nominal capacity: A measure of the charge stored by a battery (Ah for a specified C-value).
- Coulometric capacity: The total amp-hours available when the battery is discharged at a certain discharge current (electrical charge generated during electrolysis). At the specified discharge current (corresponding to the number *C*) from 100% of the charge state to the cutoff voltage. Capacity is calculated by multiplying the discharge current (in amperes) by the discharge time (in hours) and decreasing with the increase in the *C* value of the current.
- Energy or nominal energy (Wh (for a specific *C*-value)): The "capacity" of the battery, or the number of hours available when the battery is discharged at specified discharge current (corresponding to the *C* number value) from 100% of the charge state to the cutoff voltage. The energy is calculated by multiplying the discharge power (in watts) by the discharge time (in hours). Like capacity, energy decreases with increasing *C*-current.
- Lifetime (number of cycles for a specified DOD): Number of cycles of discharge; battery charging may occur before it continues to meet certain performance criteria. Lifetime is estimated for certain loading and unloading conditions. Actual battery life depends on speed and depth of cycles and other conditions such as temperature and humidity. The higher the DOD, the lower the service life.
- Specific energy (Wh/kg): Nominal battery energy per unit mass, sometimes referred to as weight energy density. Proper energy is characterized by chemical configuration adjustment and battery packing. Determines the weight of the battery required to reach a given range of electrical current.
- Specific power (W/kg): Maximum available power per unit of mass. The specific power is matched to the chemical configuration and battery pack. Specifies the weight of the battery required to reach the target performance.
- Energy density (Wh/L): The nominal battery energy per unit volume, sometimes referred to as the volume density of energy. Proper energy is characterized by chemical configuration adjustment and battery packing. Specifies the amount of battery required to reach a given range of electrical current.

- Power density (W/L): Maximum available power per unit volume. Specific power is characterized by chemistry and battery pack. It specifies the amount of battery required to reach the target performance.
- Maximum continuous discharge current: Maximum current at which the battery can be discharged continuously. This limit is usually determined by the battery manufacturer to avoid excessive discharge currents that damage the battery or reduce its capacity.
- Maximum 30-s discharge pulse current: Maximum current at which the battery can be discharged up to 30s. This border is usually described by the battery manufacturer to avoid excessive discharge currents that damage the battery or reduce its capacity.
- Charge voltage: The voltage at which the battery is charged while charging to full capacity. Circuit loading normally consists of a constant current charging until the battery voltage reaches the charging voltage, and charging with constant voltage so that the charging current reduces until the moment when it is very small.
- Float voltage: Voltage that maintains the battery, after charging up to 100% SOC, to maintain this capacity by self-discharge compensation of the battery.
- (Recommended) Charge current: Ideal current at which the battery is charged initially (to about 70% SOC) in the system constant load before proceeding to the charging at the constant voltage.
- (Maximum) Internal resistance: The electrical resistance inside the battery, generally different for charging and discharging.

REFERENCES

[1] MIT Electric Vehicle Team (Ed.), A Guide to Understanding Battery Specifications, 2008.

CHAPTER TWO

Introduction to Lithium-Sulfur Batteries

With the ever-growing global demand for energy and the extensive development of electric vehicles and portable electronic devices, progress in energy storage systems is becoming increasingly important [1–12]. Intensive use of oil for transport has a negative impact on the environment and quality of life [13–17]. Clean energy sources, such as solar and wind energy, are becoming increasingly important. The use of solar and wind energy is less profitable without energy storage, which is important for the efficient and economical storage of renewable energy to be competitive on the energy market. Therefore the efficient integration of renewable energy sources both for transport and the power grid requires the extensive infrastructure of electrical energy storage systems (EES). Due to the global increase in energy demand, EES is considered to be an essential component of both stationary and mobile energy sources. Lithium-ion (Li-ion) batteries are widely used as basic EES devices in various portable electronic devices because of their low weight and high energy storage capacities relative to other types of batteries. However, the current lithium-ion battery technology does not meet high energy and power requirements for large applications such as electric vehicles with comparable driving range to internal combustion engines (ICEs). In addition to the restrictions on the use of lithium-ion cells in electric vehicles, they also are not suitable for use in military power supplies and fixed power networks, which require a higher capacity, lower cost, and greater safety [18–22]. The main disadvantage of lithium-ion batteries lies in the fundamental chemistry of the cell, which uses transition metal compounds to store electricity through topotactic reactions (inside crystal lattices) on both electrodes. The theoretical capacity of the lithium-ion battery is less than $300\,mAh\,g^{-1}$ for each known system (see Table 1).

The conventional Li-ion batteries (LIBs) based on intercalation compounds are already facing their energy density limit, so many research studies have been aimed at developing new energy-storage systems with high-energy density [23]. The next-generation lithium-sulfur (Li-S) batteries with high theoretical specific energy ($2600\,Wh\,kg^{-1}$) have been intensively

Next-generation Batteries with Sulfur Cathodes
https://doi.org/10.1016/B978-0-12-816392-4.00002-5

Table 1 Comparison of Cathode Materials for Lithium Batteries

		Capacity ($mAh\,g^{-1}$)	
Cathode Material	**Nominal Voltage (V)**	**Theoretical**	**Practical**
$LiCoO_2$	3.6	274	110–140
$LiNi_{1/3}Co_{1/3}Mn_{1/3}O_2$	4.1	278	210
$LiMn_2O_4$	3.7	148	110–120
$LiFePO_4$	3.6	170	150
S	2.2	1672	–

investigated recently due to the advantages of the cathode material sulfur, such as its high capacity (1675 $mAh\,g^{-1}$), low cost, widespread sources, and nontoxicity [24–28]. The use of Li-S batteries is limited by the low sulfur utilization and poor cycle life, which results from the poor conductivities of sulfur and its discharge products Li_2S, the shuttling of the soluble intermediates (Li_2S_n, $4 \leq n \leq 8$) between the two electrodes, and the large volumetric change (~80%) between sulfur and Li_2S [24–28].

Lots of attention is being devoted to lithium-sulfur batteries, because they can provide an energy density three to five times higher than that of the lithium-ion batteries [29]. The typical Li-S cell shown in Fig. 1 uses composite carbon-sulfur as a cathode and metallic lithium as the anode with a liquid organic electrolyte between them [30]. The scheme of configuration of a Li-S battery employing organic liquid electrolyte is shown in Fig. 1. During discharging, the sulfur is electrochemically reduced to Li_2S on the electrode through a complex process with a series of intermediate polysulfides.

A typical discharge-charge profile of a Li-S cell is shown in Fig. 2. Active sulfur is electrochemically reduced by gradual sequences of a number of intermediate polysulfides as Li_2S_x ($x=2$–8) on the surface of the electrode, of which long chain Li_2S_x polysulfides ($x=4$–8) showed very good solubility in the electrolyte and short chain Li_2S_x polysulfides ($x=2$–4) were less soluble [31,32].

Sulfur is a yellow solid nonmetal, a cyclic molecule consisting of eight atoms, called S_8. Sulfur has more than 30 different alotropic varieties [33], but the most thermodynamically stable at room temperature (RT) is alpha-bromide sulfur (α-S_8), having a molecular weight of 32.066 $g\,mol^{-1}$, having a density of 2.07 $g\,cm^{-3}$. Sulfur has a relatively low melting point of 115°C and can easily be sublimated. Rhombic α-sulfur is used in the production of sulfur electrodes. Another alotropic form, monoclinic

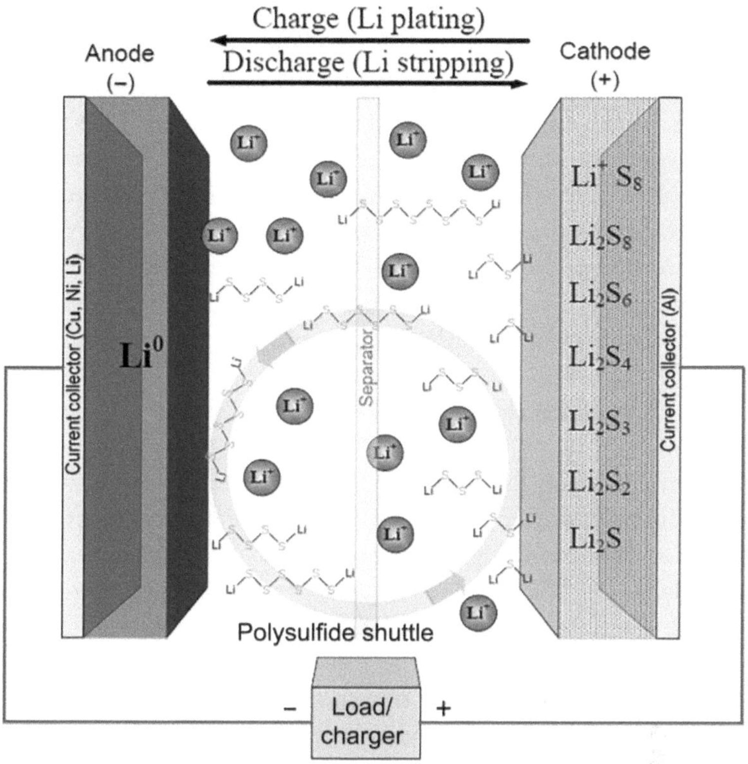

Fig. 1 Configuration Li-S battery employing organic liquid electrolyte [30].

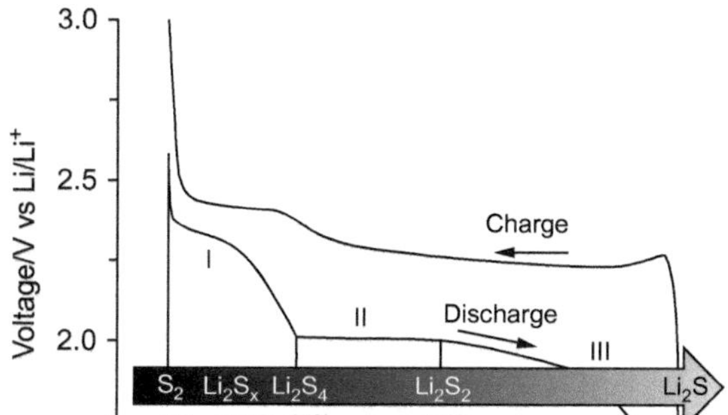

Fig. 2 Typical discharge-charge profiles of a Li-S cell, illustrating regions (I) conversion of solid sulfur to soluble polysulfides; (II) conversion of polysulfides to solid Li_2S_2; (III) conversion of solid Li_2S_2 to solid Li_2S [31,32].

beta-sulfur (β-S8) is more well known for its stability at temperatures higher than 95.5°C, and can occur when slow cooling of the molten sulfur [34–36]. Recent reports have pointed to the atypical formation of this allotropic sulfur in the Li/S system at the end of charging [37–39] or as a starting material for the positive electrode (obtained by infiltration of elemental sulfur into the CNT structure of carbon nanotubes [40]). According to [41], sulfur, with its theoretical capacity of $1675\,\mathrm{mAh\,g^{-1}}$, is the most promising alternative cathode material with high energy storage capacity. The lithium-sulfur battery (Li-S) system uses conversion chemistry (1) instead of the topotactic reaction [42]:

$$S_8 + 16Li \leftrightarrow 8Li_2S \left(E^0 = 2.2\,\mathrm{V\,vs\,Li^+/Li^0}\right) \tag{1}$$

With this reaction, each sulfur atom accepts two lithium atoms without the need for additional atoms to preserve the crystalline structure that is required for lithium-ion batteries using transition metal oxides or phosphates as cathode materials.

From the simple electrochemical reaction described by Eq. (1), every Li-S battery atom contributes to the storage of electricity. Consequently, for the same number of electrons transferred in electrochemical reactions, the weight of the active substances in the Li-S battery is significantly reduced. Although Li-S batteries have electromotive force (EMF) equal to approximately two-thirds of that offered by conventional cathode materials, sulfur can achieve a much higher energy density of $2500\,\mathrm{Wh\,kg^{-1}}$ ($2800\,\mathrm{Wh\,L^{-1}}$), assuming total reduction of elementary S to Li_2S [31,43–45]. Furthermore, sulfur is abundant in nature, is very cheap and nontoxic. For all these intriguing features, the Li-S system was one of the first intensively tested auxiliary batteries. The Li-S battery concept was introduced in 1962 [46]. A few years later, the first cell was developed with elemental sulfur as a positive electrode (cathode), lithium as negative electrode (anode), and lithium salts dissolved in organic solvents as electrolytes [45]. Most Li-S battery research was conducted between 1970 and 1980. During this period, a rich understanding of Li-S battery chemistry was gained. Nevertheless, Li-S were only marginally considered for energy storage due to their poor cyclical capacity. The low internal ionic and electronic conductivity of elemental sulfur and its final discharge products impairs the reversibility of electrochemical reactions on the cathode [45,47–49]. In addition, soluble long chain polysulfate forms can cause "chemical short circuit" of the electrochemical cell by the polysulfide shuttle-effect, a well-documented phenomena occurring in Li-S cells when using liquid electrolyte [43,47,50,51] [52]. The polysulfide shuttle-effect reduces the use of sulfur and Coulomb efficiency [43,51] and the insolubility of Li_2S and/or Li_2S_2 results in the precipitation of solids both on the cathode [49,52] and on the anode [44,53], which causes both electrodes to be electrochemically inaccessible, resulting in capacity loss [49,52]. All of these problems have contributed to stopping the commercialization of Li-S batteries. Sulfur litigation can be briefly described in the following four processes (2)–(5):

$$S_8 + 2e^- \rightarrow {S_8}^{2-} \tag{2}$$

$$ {S_8}^{2-} + 2e^- \rightarrow 2{S_4}^{2-} \tag{3}$$

$$ (2x+2y)\,Li^+ + 0.25\,(2x+y)\,{S_4}^{2-} + 0.5\,(2x+3y)\,e^- \rightarrow xLi_2S_2 + yLi_2S \tag{4}$$

$$ 2Li + Li_2S_2 + 2e \rightarrow 2Li_2S \tag{5}$$

The crown-like solid ring S_8 is first electrochemically reduced to a highly soluble S_8^{2-} by a two-stage solid and liquid reaction, showing a plateau at a voltage of approximately 2.3 V. The dissolved S_8^{2-} is then reduced to the lower order S_4^{2-} on the cathode surface, along with a number of chemical or electrochemical intermediaries, such as S_6^{2-}, S_3^{2-}, S_3^-, etc. [54,55]. This process causes a rapid increase in the viscosity of the electrolyte due to an increase in the concentration of polysilicon anions (PS) and results in a steep drop in voltage up to the lower peak observed when the solution reaches the maximum viscosity, as shown in region 2. The third process that contributes a significant portion of the cell capacity to Li-S shows a long plateau with a lower potential of ~2.1 V, which corresponds to a reduction of the biphasic solid dissolved low soluble PS to practically insoluble Li_2S_2 or Li_2S, as described in Eq. (4). The next reduction from Li_2S_2 to Li_2S takes place via a single-phase solid-solid reaction. This process has problems with weak kinetics and high polarization due to the slowdown of solid state ion diffusivity and the nature of electronic isolation of Li_2S_2 and Li_2S [53,56,57]. Despite the significant benefits mentioned previously, Li-S batteries continue to face difficulties in their practical application:

(a) the nature of the electronic and ionic sulfur and its discharge products deteriorate the use of sulfur;

(b) the dissolution of polysulfides—mediators for the cathodic reaction in a conventional liquid organic electrolyte—leads to a so-called "shuttle effect" and leads to significant loss of active cathode material and lithium corrosion on the anode;

(c) a noticeable 76% change in volume from S to Li_2S leads to destabilization of the cathode structure;

(d) adoption of a metallic lithium anode results in weakening of potential safety due to the formation of lithium dendrites and flammability in the liquid organic electrolyte.

Many efforts have been devoted to the development of new sulfur cathodes to improve electronic conductivity and suppress dissolution [25,26,31,57–63]. These papers typically focus on the design of a conductive porous matrix, such as nanostructured carbon and conductive polymers, as a host for active sulfur forms, as well as physical or chemical inhibiting dissolution and diffusion of PS to alleviate the loss of active material and suppress the shuttle effect. In addition to the development of cathodes in Li-S batteries, intensive research is also being conducted on Li-S electrolytes due to their particular and critical role. The basic function of the electrolyte for Li-S batteries is the efficient transport of Li^+ ions between the electrodes. This generally requires sufficiently high Li^+ conductivity under assumed physical, chemical, and electrochemical stability under operating conditions such as temperature and pressure as well as operating voltage windows. In addition, the electrolyte has a decisive influence on the electrode reaction mechanisms and the behavior of active sulfur and its discharge products.

According to [64], the Li-S cell has safety and protection needs that exceed those of lithium-ion batteries, as well as requiring a robust housing structure, reducing the energy density of the battery pack. The Li-S cell holds promise for the future, but the current state of the cell's degradation characteristics prevents it from competing with lithium-ion cells.

REFERENCES

[1] L. Qie, W.-M. Chen, Z.-H. Wang, Q.-G. Shao, X. Li, L.-X. Yuan, X.-L. Hu, W.-X. Zhang, Y.-H. Huang, Nitrogen-doped porous carbon nanofiber webs as a nodes for lithium ion batteries with a superhigh capacity and rate capability, Adv. Mater. 24 (15) (2012) 2047–2050.

[2] Y. Ge, H. Jiang, K. Fu, C. Zhang, J. Zhu, C. Chen, Y. Lu, Y. Qiu, X. Zhang, Copper-doped $Li_4Ti_5O_{12}$/carbon nanofiber composites as anode for high-performance sodium-ion batteries, J. Power Sources 272 (2014) 860–865.

[3] Y. Chen, X. Li, K. Park, J. Song, J. Hong, L. Zhou, Y.-W. Mai, H. Huang, J.B. Goodenough, Hollow carbon-nanotube/carbon-nanofiber hybrid anodes for Li-ion batteries, J. Am. Chem. Soc. 135 (44) (2013) 16280–16283.

[4] L. Ji, Z. Lin, M. Alcoutlabi, X. Zhang, Recent developments in nanostructured anode materials for rechargeable lithium-ion batteries, Energy Environ. Sci. 4 (8) (2011) 2682–2699.

[5] M. Gu, Y. Li, X. Li, S. Hu, X. Zhang, W. Xu, S. Thevuthasan, D.R. Baer, J.-G. Zhang, J. Liu, C. Wang, In situ TEM study of lithiation behavior of silicon

nanoparticles attached to and embedded in a carbon matrix, ACS Nano 6 (9) (2012) 8439–8447.
[6] L. Ji, X. Zhang, Fabrication of porous carbon nanofibers and their application as anode materials for rechargeable lithium-ion batteries, Nanotechnology 20 (15) (2009) 155705.
[7] X.-W. Zhang, P.K. Patil, C. Wang, A.J. Appleby, F.E. Little, D.L. Cocke, Electrochemical performance of lithium ion battery, nano-silicon-based, disordered carbon composite anodes with different microstructures, J. Power Sources 125 (2) (2004) 206–213.
[8] L. Ji, X. Zhang, Electrospun carbon nanofibers containing silicon particles as an energy-storage medium, Carbon 47 (14) (2009) 3219–3226.
[9] L. Ji, Z. Lin, A.J. Medford, X. Zhang, Porous carbon nanofibers from electrospun polyacrylonitrile/SiO_2 composites as an energy storage material, Carbon 47 (14) (2009) 3346–3354.
[10] L. Ji, X. Zhang, Evaluation of Si/carbon composite nanofiber-based insertion anodes for new-generation rechargeable lithium-ion batteries, Energy Environ. Sci. 3 (1) (2010) 124–129.
[11] L. Ji, X. Zhang, Manganese oxide nanoparticle-loaded porous carbon nanofibers as anode materials for high-performance lithium-ion batteries, Electrochem. Commun. 11 (4) (2009) 795–798.
[12] L. Ji, Y. Yao, O. Toprakci, Z. Lin, Y. Liang, Q. Shi, A.J. Medford, C.R. Millns, X. Zhang, Fabrication of carbon nanofiber-driven electrodes from electrospun polyacrylonitrile/polypyrrole bicomponents for high-performance rechargeable lithium-ion batteries, J. Power Sources 195 (7) (2010) 2050–2056.
[13] H. Wang, Y. Yang, Y. Liang, L.-F. Cui, H. Sanchez Casalongue, Y. Li, G. Hong, Y. Cui, H. Dai, LiMn1–xFexPO_4 nanorods grown on graphene sheets for ultrahigh-rate-performance lithium ion batteries, Angew. Chem. 123 (32) (2011) 7502–7506.
[14] L. Xiong, Y. Xu, T. Tao, J. Song, J.B. Goodenough, Excellent stability of spinel $LiMn_2O_4$-based composites for lithium ion batteries, J. Mater. Chem. 22 (47) (2012) 24563–24568.
[15] S.-C. Yin, H. Grondey, P. Strobel, H. Huang, L.F. Nazar, Charge ordering in lithium vanadium phosphates: electrode materials for lithium-ion batteries, J. Am. Chem. Soc. 125 (2) (2003) 326–327.
[16] B.L. Ellis, K.T. Lee, L.F. Nazar, Positive electrode materials for Li-ion and Li-batteries, Chem. Mater. 22 (3) (2010) 691–714.
[17] H. Wang, L.-F. Cui, Y. Yang, H. Sanchez Casalongue, J.T. Robinson, Y. Liang, Y. Cui, H. Dai, Mn_3O_4–graphene hybrid as a high-capacity anode material for lithium ion batteries, J. Am. Chem. Soc. 132 (40) (2010) 13978–13980.
[18] J.M. Tarascon, M. Armand, Issues and challenges facing rechargeable lithium batteries, Nature 414 (2001) 359–367, https://doi.org/10.1038/35104644.
[19] H. Li, Z. Wang, L. Chen, X. Huang, Research on advanced materials for Li-ion batteries, Adv. Mater. 21 (2009) 4593–4607, https://doi.org/10.1002/adma.200901710.
[20] B.L. Ellis, K.T. Lee, L.F. Nazar, Positive electrode materials for Li-ion and Li-batteries. Chem. Mater. 22 (2010) 691–714, https://doi.org/10.1021/cm902696j.
[21] J.B. Goodenough, Y. Kim, Challenges for rechargeable Li batteries, Chem. Mater. 22 (2010) 587–603, https://doi.org/10.1021/cm901452z.
[22] V. Etacheri, R. Marom, R. Elazari, G. Salitra, D. Aurbach, Challenges in the development of advanced Li-ion batteries: a review, Energy Environ. Sci. 4 (2011) 3243, https://doi.org/10.1039/c1ee01598b.

[23] P.G. Bruce, S.A. Freunberger, L.J. Hardwick, J.M. Tarascon, Li-O_2 and Li-S batteries with high energy storage, Nat. Mater. 11 (2012) 19–29.
[24] Y. Yang, G.Y. Zheng, Y. Cui, Nanostructured sulfur cathodes, Chem. Soc. Rev. 42 (2013) 3018–3032.
[25] Y.X. Yin, S. Xin, Y.G. Guo, L.J. Wan, Lithium-sulfur batteries: electrochemistry, materials, and prospects, Angew. Chem. Int. Ed. 52 (2013) 13186–13200.
[26] A. Manthiram, Y.Z. Fu, S.H. Chung, C.X. Zu, Y.S. Su, Rechargeable lithium-sulfur batteries, Chem. Rev. 114 (2014) 11751–11787.
[27] Y. Son, J.S. Lee, Y. Son, J.H. Jang, J. Cho, Recent advances in lithium sulfide cathode materials and their use in lithium sulfur batteries, Adv. Energy Mater. 5 (2015) 1500110.
[28] Q. Pang, X. Liang, C.Y. Kwok, L.F. Nazar, Advances in lithium-sulfur batteries based on multifunctional cathodes and electrolytes, Nat. Energy 1 (2016) 16132.
[29] N.-S. Choi, Z. Chen, S.A. Freunberger, X. Ji, Y.-K. Sun, K. Amine, G. Yushin, L.F. Nazar, J. Cho, P.G. Bruce, Challenges facing lithium batteries and electrical double-layer capacitors, Angew. Chem. Int. Ed. 51 (40) (2012) 9994–10024.
[30] Y. Mikhaylik, I. Kovalev, R. Schock, K. Kumaresan, J. Xu, J. Affinito, High energy rechargeable Li-S cells for EV application. Status, remaining problems and solutions. ECS Trans. 25 (2010) 23–34, https://doi.org/10.1149/1.3414001.
[31] X. Ji, L.F. Nazar, Advances in Li-S batteries. J. Mater. Chem. 20 (2010) 9821, https://doi.org/10.1039/b925751a.
[32] Y. Yang, G. Zheng, Y. Cui, A membrane-free lithium/polysulfide semi-liquid battery for large-scale energy storage, Energy Environ. Sci. 6 (2013) 1552, https://doi.org/10.1039/c3ee00072a.
[33] N.N. Greenwood, A. Earnshaw, Chemistry of the Elements, second ed., Butterworth-Heinemann, Oxford, 1997.
[34] L.K. Templeton, D.H. Templeton, A. Zalkin, Crystal structure of monoclinic sulfur, Inorg. Chem. 15 (1976) 1999–2001.
[35] C. Pastorino, Z. Gamba, Toward an anisotropic atom–atom model for the crystalline phases of the molecular S8 compound, J. Chem. Phys. 115 (2001) 9421–9426.
[36] C. Pastorino, Z. Gamba, Test of a simple and flexible molecule model for α-, β-, and γ-S8 crystals, J. Chem. Phys. 112 (2000) 282–286.
[37] S. Walus, C. Barchasz, J.F. Colin, J.F. Martin, E. Elkaim, J.C. Lepretre, F. Alloin, New insight into the working mechanism of lithium-sulfur batteries: in situ and operando X-ray diffraction characterization, Chem. Commun. 49 (2013) 7899–7901.
[38] J. Kulisch, H. Sommer, T. Brezesinski, J. Janek, Simple cathode design for Li-S batteries: cell performance and mechanistic insights by in operando X-ray diffraction, Phys. Chem. Chem. Phys. 16 (2014) 18765–18771.
[39] C. Villevieille, P. Novák, A metastable β-sulfur phase stabilized at room temperature during cycling of high efficiency carbon fibre–sulfur composites for Li–S batteries, J. Mater. Chem. A 1 (2013) 13089.
[40] S. Moon, Y.H. Jung, W.K. Jung, D.S. Jung, J.W. Choi, K. Kim do, Encapsulated monoclinic sulfur for stable cycling of Li-S rechargeable batteries, Adv. Mater. 25 (2013) 6547–6553.
[41] Z. Liu, W. Fu, C. Liang, Lithium–sulfur batteries, in: C. Daniel, J.O. Besenhard (Eds.), Handbook of Battery Materials, second ed., Wiley-VCH Verlag GmbH & Co. KGaA, 2011. Chapter 24.
[42] X.L. Ji, K.T. Lee, L.F. Nazar, A highly ordered nanostructured carbon–sulphur cathode for lithium–sulphur batteries, Nat. Mater. 8 (2009) 500–506.
[43] Y.V. Mikhaylik, J.R. Akridge, Polysulfide shuttle study in the Li/S battery system, J. Electrochem. Soc. 151 (2004) A1969–A1976.
[44] H. Yamin, E. Peled, Electrochemistry of a nonaqueous lithium/sulfur cell, J. Power Sources 9 (1983) 281–287.

[45] M.L.B. Rao, Organic electrolyte cells, US patent 3,413,154, 1968.
[46] D. Herbert, J. Ulam, Electric dry cells and storage batteries, US patent 3,043, 896, 1962.
[47] J. Shim, K.A. Striebel, E.J. Cairns, The lithium/sulfur rechargeable cell. Effects of electrode composition and solvent on cell performance, J. Electrochem. Soc. 149 (2002) A1321–A1325.
[48] D. Marmorstein, T.H. Yu, K.A. Striebel, F.R. McLarnon, J. Hou, E.J. Cairns, Electrochemical performance of lithium/sulfur cells with three different polymer electrolytes, J. Power Sources 89 (2000) 219–226.
[49] B.H. Jeon, J.H. Yeon, K.M. Kim, I.J. Chung, Preparation and electrochemical properties of lithium–sulfur polymer batteries, J. Power Sources 109 (2002) 89–97.
[50] B.M.L. Rao, J.A. Shropshire, Effect of sulfur impurities on Li / TiS_2 cells, J. Electrochem. Soc. 128 (1981) 942–945.
[51] Y.V. Mikhaylik, J.R. Akridge, Low temperature performance of Li/S batteries, J. Electrochem. Soc. 150 (2003) A306–A311.
[52] S.E. Cheon, K.S. Ko, J.H. Cho, S.W. Kim, E.Y. Chin, H.T. Kim, Rechargeable lithium sulfur battery I. Structural change of sulfur cathode during discharge and charge, J. Electrochem. Soc. 150 (2003) A796–A799.
[53] J.I. Yamaki, S.I. Tobishima, Y. Sakurai, K.I. Saito, K. Hayashi, Safety evaluation of rechargeable cells with lithium metal anodes and amorphous V_2O_5 cathodes, J. Appl. Electrochem. 28 (1998) 135–140.
[54] C. Barchasz, F. Molton, C. Duboc, J.C. Lepretre, S. Patoux, F. Alloin, Lithium/sulfur cell discharge mechanism: an original approach for intermediate species identification, Anal. Chem. 84 (2012) 3973–3980, https://doi.org/10.1021/ac2032244.
[55] S.S. Zhang, J.A. Read, A new direction for the performance improvement of rechargeable lithium/sulfur batteries, J. Power Sources 200 (2012) 77–82, https://doi.org/10.1016/j.jpowsour.2011.10.076.
[56] H. Yamin, E. Peled, Electrochemistry of a non-aqueous lithium sulfur cell, J. Power Sources 9 (1983) 281–287, https://doi.org/10.1016/0378-7753(83)87029-3.
[57] M. Barghamadi, A. Kapoor, C. Wen, A review on Li-S batteries as a high efficiency rechargeable lithium battery, J. Electrochem. Soc. 160 (2013) A1256–A1263, https://doi.org/10.1149/2.096308jes.
[58] D. Bresser, S. Passerini, B. Scrosati, Recent progress and remaining challenges in sulfur-based lithium secondary batteries—a review, Chem. Commun. (Camb.) 49 (2013) 10545–10562, https://doi.org/10.1039/c3cc46131a.
[59] S. Evers, L.F. Nazar, New approaches for high energy density lithium-sulfur battery cathodes, Acc. Chem. Res. 46 (2013) 1135–1143, https://doi.org/10.1021/ar3001348.
[60] A. Manthiram, Y. Fu, Y.S. Su, Challenges and prospects of lithium-sulfur batteries, Acc. Chem. Res. 46 (2013) 1125–1134, https://doi.org/10.1021/ar300179v.
[61] M.K. Song, E.J. Cairns, Y. Zhang, Lithium/sulfur batteries with high specific energy: old challenges and new opportunities, Nanoscale 5 (2013) 2186–2204, https://doi.org/10.1039/c2nr33044j.
[62] L. Chen, L.L. Shaw, Recent advances in lithium-sulfur batteries, J. Power Sources 267 (2014) 770–783, https://doi.org/10.1016/j.jpowsour.2014.05.111.
[63] G.Y. Xu, B. Ding, J. Pan, P. Nie, L.F. Shen, X.G. Zhang, High performance lithium-sulfur batteries: advances and challenges, J. Mater. Chem. A 2 (2014) 12662–12676, https://doi.org/10.1039/C4ta02097a.
[64] C.E. Parfitt, Characterisation, Modelling and Management of Lithium-Sulphur Batteries for Spacecraft Applications (Ph.D. thesis), University of Warwick, 2012.

CHAPTER THREE

Mechanisms of Shuttle Effect and Loss of Capacity

According to Liu et al. [1] polysulfide shuttle effect is a phenomenon characteristic of Li-S batteries with liquid electrolytes. This phenomenon is responsible for high self-discharge, low coulombic efficiency, significant sulfur migration, and rapid decrease of battery capacity. Polysulfide shuttle effect gives high mobility to sulfur forms and leads to uncontrolled precipitation of Li_2S_2 and/or Li_2S. High mobility of the sulfur form significantly affects the Li-S cells charging-discharging profiles. The migration of sulfur leads to self-discharge of the cell. Precipitation of the sulfide forms outside the cathode region results in the loss of active substances and structural cell damage [2–4].

Recently, various attempts have been made to limit the impact of polisulfide shuttle effect.

For example, as described in [5], a TiO_2-nitrogen doped graphene/sulfur (TiO_2-NG/S) hybrid structure was obtained using the atomic layer deposition (ALD) method. The nitrogen doped graphene was used as a conductive matrix and the TiO_2 coating on the NG/S electrodes surface was used to inhibit lithium polysulfide shuttle.

During studies described in [6], a TiO_2 nanowire-embedded graphene (TiO_2 NW/G) hybrid membrane was used. In this membrane, the graphene with high conductivity was used as the current collector, while the TiO_2 NWs were used as a polysulfides shuttling inhibitor as well as the catalyst to accelerate the polysulfide reduction and oxidation.

According to [7], a carbonized bacterial cellulose/titania (CBC/TiO_2) modified separator was elaborated. This separator restrained the shuttle effect of Li-S cells with strong physical and chemical adsorption of polysulfides.

During studies described in [8], a free-standing MnO_2 nanowires and graphene nanoscroll (GNSM) interlayer was synthesized, where the weight ratio of graphene nanoscroll to MnO_2 nanowires was 4:1. The insertion of the hybrid GNSM interlayer between sulfur cathode and separator reduced the overall electrode resistance. It also offered strong physical/chemical

Next-Generation Batteries With Sulfur Cathodes
https://doi.org/10.1016/B978-0-12-816392-4.00003-7

interactions for efficiently mitigating the shuttling of polysulfides, ensuring continuous reactivation and reutilization of the trapped sulfur active materials.

According to [9], a V_2O_5-decorated carbon nanofiber interlayer for suppressing self-discharge and shuttle effect was developed.

3.1 ORIGIN OF POLYSULFIDE SHUTTLE-EFFECT

The polysulfide shuttle effect is a concentration-dependent diffusion phenomenon in the liquid phase together with redox reactions in the form of sulfur and sulfides that have oxidation states in the range from −2 to 0. The Li-S battery uses liquid base electrolytes, which have a high solubility of polysulfides to improve the maximum strength currents and allow the cell to work within a wide temperature range [10–14]. However, solubility and motility of polysulfides in organic electrolytes results in the phenomenon of shuttle effect during loading [2, 10, 15–17]. The elemental sulfur is first discharged, and two plates appear on the discharge curve at room temperature [10, 13, 15, 16]. During this discharging, elemental sulfur (S_8) particles are reduced to polysulfides (Li_2S_x). The high voltage plateau (2.3–2.4 V) corresponds to the formation of high-order polysulfides (Li_2S_h, $h > 4$), taking 0.5 electrons per sulfur atom. The corresponding sulfur specific capacity is 419 mAh g^{-1}, according to Eq. (6). Further reduction of high-order polysulfides to low-order polysulfides (Li_2S_l, $l < 4$) occurs for the low voltage plateau (~2.1 V) according to Eq. (7). One additional electron is taken on a sulfur atom for a low voltage plateau, which corresponds to a specific capacity of 837 mAh g^{-1}.

Due to the low solubility and slow kinetics of lithium disulfide in the electrolyte, the reduction to Li_2S cannot be completed. Therefore, the total specific capacity for the high and low voltage plateau is 1256 mAh g^{-1} based on the mass of sulfur in DOL/DME (1,3-dioxolane/dimethoxyethane) [15], which is similar to that obtained in the electrolyte based on tetrahydrofuran (THF) at C/50 discharge current [13].

$$S_8 + 4e^- = 2{S_4}^{2-} \quad (6)$$

$$ {S_4}^{2-} + 4e^- = 2S^{2-} + {S_2}^{2-} \quad (7)$$

Charging of the Li-S cell is a multiple of the electron transfer process, low-order polysulfides not directly oxidized to elemental sulfur. Instead, high-order polysulfides (${S_h}^{2-}$) are created at the cathode (Eq. 8) and a concentration gradient is formed in the electrochemical cell. The S_h^{2-} concentration is higher at the cathode than at the anode. These high-order polysulfides diffuse into the anode and react chemically with lithium to form lower-order polysulfides (S_{h-x}^{2-}, Eq. 9) if the lithium electrode is not protected. As a result of diffusion and the S_h^{2-} reaction, there is a high concentration of lower-order polysulfides on the anode side (S_{h-x}^{2-}). These lower-order polysulfides diffuse back into the cathode and are electrochemically oxidized to reform higher-order polysulfides (Eq. 10). Polysulfide diffusion between the two Li electrodes transfers the sulfur back and forth.

This phenomenon is known as the polysulfide shuttle effect. Fig. 3 shows such a transfer mechanism [2, 10, 15–17]. Chemical reactions of the polysulfide shuttle effect are described by the following equations:

Electrochemical oxidation of sulfide and disulfide to higher-order polysulfides at the cathode:

$$2S^{2-} + S_2^{2-} = S_h^{2-} + xe^- \tag{8}$$

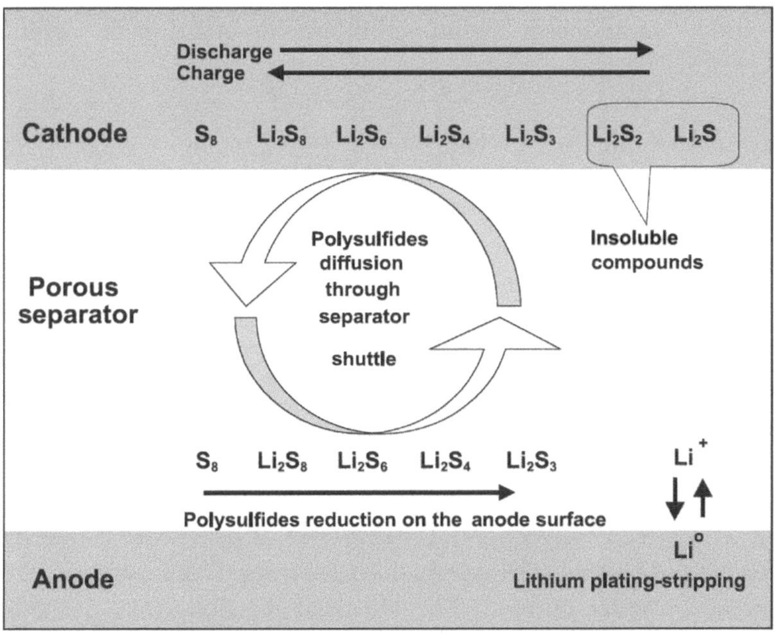

Fig. 3 Operation of Li-S batteries and the phenomenon of shuttle effect. *Taken from J.R. Akridge, Y.V. Mikhaylik, N. White, Li/S fundamental chemistry and application to high-performance rechargeable batteries, Solid State Ionics 175 (2004) 243–245.*

Chemical reduction of high-order polysulfides that diffuse into the anode:

$$(h-x)S_h{}^{2-} + 2x\mathrm{Li}^0 = hS_{h-x}{}^{2-} + 2x\mathrm{Li}^+ \quad (9)$$

Electrochemical oxidation of low-order polysulfides, which diffuse back to the cathode:

$$hS_{h-x}{}^{2-} = (h-x)S_h{}^{2-} + 2xe^- \quad (10)$$

3.2 INFLUENCE OF SHUTTLE-EFFECT ON CHARGING PROFILE

The polysulfide shuttle effect phenomenon is associated with higher-order polysulfides, due to their high solubility and mobility, and the reduction of charging efficiency mainly for the high voltage plateau. The efficiency of charging depends on the charging current and constant transfer. In order to discuss the effects of the polysulfide shuttle effect, the equation of polysulfide dynamics is introduced, which is expressed by Eq. (11) [15]. The transfer constant (k_s) depends on the chemical properties and concentration of electrolyte components, including lithium salts and organic solvents.

$$d[S_h]/dt = i/q_h - k_s[S_h] \quad (11)$$

In this equation, $[S_h]$ is the concentration of higher-order polysulfides, t is the time, and i is the current density, and q_h is the proper capacity at the high voltage plateau. The cell can be fully charged at high charging current or low solid transfer. However, if the charging current is low and the transfer constant is high, the cell may not be fully charged, even after an infinite charging time. In this scenario, electricity, instead of being accumulated, is consumed to heat the cell, due to the chemical reaction of the polysulfides with the lithium electrode. This situation is also known as the chemical short circuit of the Li-S cell. The heat generated in the cell is discharged into the environment, and the cell reaches equilibrium at the relatively low temperature of the cell. More than 15 min is required for the cell to achieve these conditions with forced convection [10]. However, if heat dissipation is halted, cell voltage and heat release increase due to increased concentration of polysulfides and transfer current. The chemical reaction heats the cell and increases the transfer constant. When the transfer current increases to a value greater than the charging current, the concentration of polysulfides is reduced in the cathode, thus lowering the cell voltage. Finally, the cell reaches a different equilibrium with equal transfer and charging currents. In other words, the transfer provides effective overload protection [10, 15].

3.3 EFFECT OF SHUTTLE EFFECT ON CHARGE-DISCHARGE CAPACITY

The chemical reduction of the transferred polysulfides is always present in Li-S batteries. The electrochemical reaction competes with the chemical reaction in the cell; therefore, the total cell capacities in both charge and discharge cycles greatly affect the transport of polysulfides. The transfer constant, which is closely related to the rate of chemical reactions, is a numerical measure of the impact of the shuttle effect phenomenon. Slow chemical reactions, i.e., low transfer constant, are always desirable to achieve high efficiency of energy storage [15]. When the cell is loaded and unloaded at a constant temperature, the transfer constant can be considered as fixed; the total charging and discharging capacities depend on the currents. According to Eq. (11), with a fixed transfer constant, high currents favor electrochemical reactions. Hence, coulombic efficiency can be improved by using large currents in both charge and discharge cycles. However, high current increases the temperature of the cell, and thus increases the constant transfer. High current does not always result in good energy efficiency. In addition, inhomogeneous precipitation of the low solubility form (sulfur in the charge and sulfide/disulfide cycle in the discharge cycle) leads to insufficient use of active materials. Improperly high charging-discharging current can cause failure or low cell capacities in both cycles. In the storage conditions on the shelf, only the capacity for the high voltage plateau disappears after a certain storage time, within a few hours to several days, for the low voltage plateau is stable for several weeks. The rapid drop in capacity for the high voltage plateau occurs due to the high reactivity of higher-order polysulfides with lithium electrodes [2, 13, 15, 16].

3.4 THE MECHANISM OF CAPACITY LOSS

A rapid decrease in the capacity of Li-S batteries is always observed. The main reason for this is the high mobility of polysulfide forms in liquid electrolytes. The loss of capacitance in the high voltage plateau is mainly attributed to the phenomenon of the polysulfide shuttle effect, whereas in the low voltage plateau it is associated with the failure of the electrode structure due to successive Li_2S precipitates on the surfaces of both electrodes

[2–4, 18–20]. At the cathode, Li_2S is formed by electrochemical reduction, and on the anode, Li_2S results from the chemical reduction of polysulfides that diffuse from the cathode. The solid Li_2S film was observed on the surface of the cathode, and this film serves as a passivation layer. Even when the battery is fully charged, the Li_2S passivation layer still remains on the cathode carbon matrix [2]. This passivation layer leads not only to the loss of capacitance, but also to the high resistance of the cell, and thus leads to high polarity of the electrode [2, 3]. Stresses generated during cyclic battery operation can also lead to cell damage through the development of microcracks or cracks. A similar Li_2S passivation layer is also formed on the surface of the anode, due to the surface reaction of polysulfides with metallic lithium. The formation of the passivation layer on the anode side can protect the cyclic work of the metallic lithium; however, the capacity of the cell decreases and the cell resistance increases [2, 3].

REFERENCES

[1] Z. Liu, W. Fu, C. Liang, Chapter 24 Lithium–sulfur batteries, in: C. Daniel, J.O. Besenhard (Eds.), Handbook of Battery Materials, second ed., Wiley-VCH Verlag GmbH & Co. KGaA, 2011.

[2] S.E. Cheon, K.S. Ko, J.H. Cho, S.W. Kim, E.Y. Chin, H.T. Kim, Rechargeable lithium sulfur battery I. Structural change of sulfur cathode during discharge and charge, J. Electrochem. Soc. 150 (2003) A796–A799.

[3] S.E. Cheon, K.S. Ko, J.H. Cho, S.W. Kim, E.Y. Chin, H.T. Kim, Rechargeable lithium sulfur battery II. Rate capability and cycle characteristics, J. Electrochem. Soc. 150 (2003) A800–A805.

[4] S.E. Cheon, S.S. Choi, J.S. Han, Y.S. Choi, B.H. Jung, H.S. Lim, Capacity fading mechanisms on cycling a high-capacity secondary sulfur cathode, J. Electrochem. Soc. 151 (2004) A2067–A2073.

[5] M. Yu, J. Ma, H. Song, A. Wang, F. Tian, Y. Wang, H. Qiu, R. Wang, Atomic layer deposited TiO_2 on a nitrogen-doped graphene/sulfur electrode for high performance lithium–sulfur batteries, Energy Environ. Sci. 9 (4) (2016) 1495.

[6] G. Zhou, Y. Zhao, C. Zu, A. Manthiram, Free-standing TiO_2 nanowire–embedded graphene hybrid membrane for advanced Li/dissolved polysulfide batteries, Nano Energy 12 (2015) 240.

[7] F. Li, G. Wang, P. Wang, J. Yang, K. Zhang, Y. Liu, Y. Lai, High-performance lithium–sulfur batteries with a carbonized bacterial cellulose/TiO_2 modified separator, J. Electroanal. Chem. 788 (2017) 150.

[8] Y. Guo, G. Zhao, N. Wu, Y. Zhang, M. Xiang, B. Wang, H. Liu, H. Wu, Efficient synthesis of graphene nanoscrolls for fabricating sulfur-loaded cathode and flexible hybrid interlayer toward high-performance Li–S batteries, ACS Appl. Mater. Interfaces 8 (50) (2016) 34185.

[9] M. Liu, Q. Li, X. Qin, G. Liang, W. Han, D. Zhou, Y.B. He, B. Li, F. Kang, Suppressing self-discharge and shuttle effect of lithium–sulfur batteries with V_2O_5-decorated carbon nanofiber interlayer, Small 13 (12) (2017) 1602539–1602545.

[10] Y.V. Mikhaylik, J.R. Akridge, Low temperature performance of Li/S batteries, J. Electrochem. Soc. 150 (2003) A306–A311.

[11] X.M. He, W.H. Pu, J.U. Ren, L. Wang, J.L. Wang, C.Y. Jiang, C.R. Wan, Charge/discharge characteristics of sulfur composite electrode at different temperature and current density in rechargeable lithium batteries, Ionics 14 (2008) 335–337.
[12] H.S. Ryu, H.J. Ahn, K.W. Kim, J.H. Ahn, K.K. Cho, T.H. Nam, J.U. Kim, G.B. Cho, Discharge behavior of lithium/sulfur cell with TEGDME based electrolyte at low temperature, J. Power Sources 163 (2006) 201–206.
[13] R.D. Rauh, K.M. Abraham, G.F. Pearson, J.K. Surprenant, S.B. Brummer, A lithium/dissolved sulfur battery with an organic electrolyte, J. Electrochem. Soc. 126 (1979) 523–527.
[14] R.D. Rauh, F.S. Shuker, J.M. Marston, S.B. Brummer, Formation of lithium polysulfides in aprotic media, J. Inorg. Nucl. Chem. 39 (1977) 1761–1766.
[15] Y.V. Mikhaylik, J.R. Akridge, Polysulfide shuttle study in the Li/S battery system, J. Electrochem. Soc. 151 (2004) A1969–A1976.
[16] J. Shim, K.A. Striebel, E.J. Cairns, The lithium/sulfur rechargeable cell. Effects of electrode composition and solvent on cell performance, J. Electrochem. Soc. 149 (2002) A1321–A1325.
[17] B.M.L. Rao, J.A. Shropshire, Effect of sulfur impurities on Li/TiS_2 cells, J. Electrochem. Soc. 128 (1981) 942–945.
[18] H. Yamin, E. Peled, Electrochemistry of a nonaqueous lithium/sulfur cell, J. Power Sources 9 (1983) 281–287.
[19] S.E. Cheon, J.H. Cho, K.S. Ko, C.W. Kwon, D.R. Chang, H.T. Kim, S.W. Kim, Structural factors of sulfur cathodes with poly (ethylene oxide) binder for performance of rechargeable lithium sulfur batteries, J. Electrochem. Soc. 149 (2002) A1437–A1441.
[20] R. Elazari, G. Salitra, Y. Talyosef, J. Grinblat, C. Scordilis-Kelley, A. Xiao, J. Affinito, D. Aurbach, Morphological and structural studies of composite sulfur electrodes upon cycling by HRTEM, AFM and Raman spectroscopy, J. Electrochem. Soc. 157 (2010) A1131–A1138.

FURTHER READING

[21] J.R. Akridge, Y.V. Mikhaylik, N. White, Li/S fundamental chemistry and application to high-performance rechargeable batteries, Solid State Ionics 175 (2004) 243–245.

CHAPTER FOUR

Cell Configuration

The lithium-sulfur system, because of its high volumetric and gravimetric energy densities, has attracted interest both for secondary and primary battery applications [1].

According to [2], lithium-sulfur primary batteries are seldom reported in the existing literature. However, one exemplary configuration has been described. A cauliflower-like hierarchical porous C/S cathode was designed for facilitating lithium-ion transport and sulfur accommodation. This type of cathode released about 1300 mAh g^{-1} (S) capacity at sulfur loading of 6–14 mg cm^{-2} and showed excellent shelf stability during a month-long test at room temperature. As a result, the assembled Li-S soft package battery achieved an energy density of 504 Wh kg^{-1} (654 Wh L^{-1}).

The other exemplary type of primary cell with a lithium anode and sulfur cathode was described in [3]. In that study it was stated that the lithium-sulfur dioxide battery is a primary battery system with many advantages over conventional batteries. Its energy density was up to 330 Wh/kg, two to four times greater than zinc batteries, and it could perform to temperatures as low as −54°C. The battery could withstand high-temperature storage at 71°C for long periods of time and its shelf life can be 5–10 years at 21°C. This Li/SO_2 system provides an all-purpose, all-climate primary battery.

In other literature, however, more attention is devoted to Li-S secondary cells and batteries. These often describe still experimental configurations and, according to the review [4], the development of novel structural configurations, including hierarchical gradient cathodes, solid-state electrolytes, functional separators and anodic protection, are crucial for further enhancement of Li-S batteries.

The configuration of Li-S battery cells is much more complex than in the case of traditional batteries using solid active substances; this is due to sulfur phase transitions occurring during battery operation. Although solid sulfur is often used as a starting material for cathode production, it dissolves in the electrolyte during discharge of the cell. Forms of condensed sulfur form a catholyte, which is the main form of sulfur during the operation of Li-S batteries. The catholyte diffusion is the beginning of the polysulfide shuttle

Next-generation Batteries with Sulfur Cathodes
https://doi.org/10.1016/B978-0-12-816392-4.00004-9

effect. To avoid the negative effects of the polysulfide shuttle effect, a number of cell configurations have been developed over the past few decades. In the initial and the simplest configuration [5], the positive electrode contains solid sulfur and soot (or other inert material with high electronic conductivity), the negative electrode is made of lithium, and the electrolyte contains a lithium salt soluble in organic solvents. Suitable binders are used to preserve the integrity of the structure [5, 6] electrodes. The observed poor discharge efficiency is attributed to the formation of soluble polysulfides that diffuse between the cathode and the anode, where self-discharge occurs [7]. The insulating material of the cathode and the Li_2S or Li_2S_2 final product are responsible for the poor efficiency of cyclic work [8]. In an attempt to use a liquid cathode [8], soluble sulfur instead of solid sulfur is used to improve discharge efficiency. A polysulfide solution (Li_2S_n) in THF containing $LiAsF_6$ is produced by the reaction of a stoichiometric amount of sulfur with Li_2S. The liquid electrode maintains a very good electrical contact with the current collector during the discharge process. The obvious advantage of a liquid cathode is the high use of active substances; 75% utilization of sulfur is obtained at very low current (C/50) and at room temperature. The slow reduction kinetics of the low-order polysulfides and the Li_2S passivation layer formed on the cathode create a low ability to withstand the maximum currents through the liquid cathode. The soluble Li_2S additives and/or complexing agents for S_n^{2-} are required to further improve the discharge rate and capacity. The first prevent the formation of a passivation layer at the cathode, and the second increases the rate of reduction of low-order polysulfides. Although there are chemical reactions between metallic lithium and polysulfides, the concentrated polysulfide solution is quite compatible with metallic lithium [8]. Li_2S formation on the anode passivates metallic lithium and gives high system compatibility. Interestingly, the increase in dendritic lithium is not observed in the Li-S cell with a simple cell configuration when the polysulfides can be transferred between the electrodes. When replacing liquid electrolytes by polymer electrolytes or gel electrolytes, the ability to operate cyclic Li-S batteries is significantly improved by slowing the polysulfide shuttle effect [9–18]. Polymer electrolytes function as both electrolyte and separator. Since polysulfide diffusion in polymeric electrolytes is much lower than in liquid electrolytes, polysulfides are limited in the cathode area during battery operation. The migration of sulfur to the anode area is avoided. Thus, the loss of active substances by diffusion does not occur in Li-S batteries using polymeric electrolytes. However, the ionic conductivity of polymeric electrolytes, e.g., poly (ethylene oxide) (PEO) [19] based electrolyte, is low at room temperature. Li-S batteries with polymer electrolytes require a high operating temperature to obtain sufficient ion conductivity. The ionic conductivity at room temperature can be improved by the addition of ceramic fillers such as Al_2O_3, ZrO_2, and others [18, 20, 21]. However, the loss of capacity is still a problem. The formation of lithium dendrites cannot be avoided if a polymeric or gel electrolyte is used. Safety issues related to the anode can be a problem in the configuration of cells with polymeric electrolytes. The configuration with only solid components is an advanced system based on glass ceramic electrolytes [22–26]. Glass ceramics such as Li_2S-SiS_2 [27] and Li_2S-P_2S_5 [28] show high conductivity at room temperature, and glass ceramic electrolytes have a lithium-ion transport number equal to one, which is a great advantage over polymeric electrolytes [22]. The electrolyte is a positive electrode element to improve the ionic conductivity of the electrode [22]. However, considering the very low ionic and electronic conductivity of sulfur and lithium sulfide, it is a huge challenge to overcome the internal problems of Li-S battery chemistry even before the configuration of all solid components.

Although the configuration with all solid components shows a better ability to work cyclically at low current, the discharging potential is only 1.4–1.5 V due to higher resistance at the interface, and this thus diminishes the advantage of energy density for the Li-SW system. Zhu et al. [29] describe a bifunctional intermediate layer conducting the ions and electrons intended for the Li-S battery with all solid elements. This intermediate layer is intercalated between the cathode and the solid electrolyte (SPE). The Li-S battery generates a gradient ion-electron change, significantly increasing compatibility at the cathode/SPE interface. The Li-S battery with an intermediate layer exhibits higher utilization of the active material and thus higher efficiency. Lopez-Aranguren et al. [30] describe a battery with all solid elements, including a hybrid metallic anode, an S/C composite cathode, and a $LiBH_4$ solid electrolyte. The hybrid anode was made of $0.8MgH_2 + 0.2TiH_2$ with improved electrochemical properties. The battery showed a high reversible specific capacity of 910 mAh g^{-1} with a discharge plateau at 1.8 and 1.4 V. Furthermore, the capacity remained at 85% of the initial value during the first 25 charging/discharging cycles.

References

[1] E. Peled, A. Gorenshtein, M. Segal, Y. Sternberg, Rechargeable lithium-sulfur battery, J. Power Sources 26 (1989) 269–271.
[2] Y. Ma, H. Zhang, B. Wu, M. Wang, X. Li, H. Zhang, Lithium sulfur primary battery with super high energy density: based on the cauliflower-like structured C/S cathode, Sci. Rep. 5 (2015), https://doi.org/10.1038/srep14949. Article number: 14949.

[3] D. Linden, B. McDonald, The lithium—sulfur dioxide primary battery—its characteristics, performance and applications, J. Power Sources 5 (1) (1980) 35–55.
[4] M. Liu, X. Qin, Y.-B. He, B. Li, F. Kang, Recent innovative configurations in high-energy lithium sulfur batteries, J. Mater. Chem. A 5 (2017) 5222.
[5] M.L.B. Rao, Organic electrolyte cells, US Patent 3,413,154, 1968.
[6] D.A. Nole, V. Moss, Battery employing lithium-sulphur electrodes with non-aqueous electrolyte, US Patent 3,532,543, 1970.
[7] G. Eichinger, J.O. Besenhard, High energy density lithium cells: part II. Cathodes and complete cells, J. Electroanal. Chem. Interfacial Electrochem. 72 (1976) 1–31.
[8] R.D. Rauh, K.M. Abraham, G.F. Pearson, J.K. Surprenant, S.B. Brummer, A lithium/dissolved sulfur battery with an organic electrolyte, J. Electrochem. Soc. 126 (1979) 523–527.
[9] D. Marmorstein, T.H. Yu, K.A. Striebel, F.R. McLarnon, J. Hou, E.J. Cairns, Electrochemical performance of lithium/sulfur cells with three different polymer electrolytes, J. Power Sources 89 (2000) 219–226.
[10] B.H. Jeon, J.H. Yeon, K.M. Kim, I.J. Chung, Preparation and electrochemical properties of lithium–sulfur polymer batteries, J. Power Sources 109 (2002) 89–97.
[11] J.H. Park, S.Y. Yeo, J.K. Park, Y.M. Lee, Electrochemical performance of lithium sulfur batteries with plasticized polymer electrolytes based on P(VdF-co-HFP), J. Korean Electrochem. Soc. 13 (2010) 110–115.
[12] H.S. Ryu, J.W. Choi, J.H. Ahn, G.B. Cho, H.J. Ahn, The electrochemical properties of poly(acrylonitrile) polymer electrolyte for Li/S battery, in: H.S. Kim, Y.B. Li, S.W. Lee (Eds.), Eco-Materials Processing and Design VII, vol. 510–511, Trans Tech Publications Ltd, Zurich-Uetikon, 2006, pp. 50–53.
[13] J. Shin, B. Jung, S. Jeong, K. Kim, H. Ahn, K. Cho, J. Ahn, Electrochemical and interfacial properties of $(PEO)_{10}LiCF_3SO_3-Al_2O_3$ nanocomposite polymer electrolytes using ball milling, Met. Mater. Int. 10 (2004) 177–183.
[14] V.S. Kolosnitsyn, E.V. Karaseva, N.A. Amineva, G.A. Batyrshina, Cycling lithium–sulfur batteries, Russ. J. Electrochem. 38 (2002) 329–331.
[15] J.H. Shin, K.W. Kim, H.J. Ahn, J.H. Ahn, Electrochemical properties and interfacial stability of $(PEO)_{10}LiCF_3SO_3$–Ti_nO_{2n-1} composite polymer electrolytes for lithium/sulfur battery, Mater. Sci. Eng., B 95 (2002) 148–156.
[16] X.J. Zhu, Z.Y. Wen, Z.H. Gu, Z.X. Lin, Electrochemical characterization and performance improvement of lithium/sulfur polymer batteries, J. Power Sources 139 (2005) 269–273.
[17] J.W. Choi, J.H. Kim, G. Cheruvally, J.H. Ahn, K.W. Kim, H.J. Ahn, J.U. Kim, Microporous poly(vinylidene fluoride-co-hexafluoropropylene) polymer electrolytes for lithium/sulfur cells, J. Ind. Eng. Chem. 12 (2006) 939–949.
[18] S.S. Jeong, Y.T. Lim, Y.J. Choi, G.B. Cho, K.W. Kim, H.J. Ahn, K.K. Cho, Electrochemical properties of lithium sulfur cells using PEO polymer electrolytes prepared under three different mixing conditions, J. Power Sources 174 (2007) 745–750.
[19] H.J. Ahn, K.W. Kim, J.H. Ahn, G. Cheruvally, Secondary Batteries–Lithium Rechargeable Systems: Lithium-Sulfur. Encyclopedia of Electrochemical Power Sources, Elsevier, 2010.
[20] J. Hassoun, B. Scrosati, Moving to a solid-state configuration: a valid approach to making lithium-sulfur batteries viable for practical applications, Adv. Mater. 22 (2010) 5198–5201.
[21] J. Hassoun, B. Scrosati, A high-performance polymer tin sulfur lithium ion battery, Angew. Chem. Int. Ed. 49 (2010) 2371–2374.
[22] A. Hayashi, T. Ohtomo, F. Mizuno, K. Tadanaga, M. Tatsumisago, All-solid-state Li/S batteries with highly conductive glass–ceramic electrolytes, Electrochem. Commun. 5 (2003) 701–705.

[23] A. Hayashi, R. Ohtsubo, T. Ohtomo, F. Mizuno, M. Tatsumisago, All-solid-state rechargeable lithium batteries with Li_2S as a positive electrode material, J. Power Sources 183 (2008) 422–426.
[24] T. Kobayashi, Y. Imade, D. Shishihara, K. Homma, M. Nagao, R. Watanabe, T. Yokoi, A. Yamada, R. Kanno, T. Tatsumi, All solid-state battery with sulfur electrode and thio-LISICON electrolyte, J. Power Sources 182 (2008) 621–625.
[25] N. Machida, K. Kobayashi, Y. Nishikawa, T. Shigematsu, Electrochemical properties of sulfur as cathode materials in a solid-state lithium battery with inorganic solid electrolytes, Solid State Ionics 175 (2004) 247–250.
[26] N. Machida, T. Shigematsu, An all-solid-state lithium battery with sulfur as positive electrode materials, Chem. Lett. 33 (2004) 376–377.
[27] A. Pradel, M. Ribes, Electrical properties of lithium conductive silicon sulfide glasses prepared by twin roller quenching, Solid State Ionics 18–19 (1986) 351–355.
[28] K. Minami, F. Mizuno, A. Hayashi, M. Tatsumisago, Lithium ion conductivity of the Li_2S–P_2S_5 glass-based electrolytes prepared by the melt quenching method, Solid State Ionics 178 (2007) 837–841.
[29] Y. Zhu, J. Li, J. Liu, A bifunctional ion-electron conducting interlayer for high energy density all-solid-state lithium-sulfur battery, J. Power Sources 351 (2017) 17–25.
[30] P. Lopez-Aranguren, N. Berti, A.H. Dao, J. Zhang, F. Cuevas, M. Latroche, C. Jordy, An all-solid-state metal hydride–sulfur lithium-ion battery, J. Power Sources 357 (2017) 56–60.

CHAPTER FIVE

Materials for Positive Electrode (Cathode)

According to [1], a positive electrode is crucial for Li-S batteries. The main difficulties for cyclic Li-S battery operation lie in the high mobility of sulfur compounds. Sulfur changes its form from solid to liquid phase when the battery is partially discharged and precipitates in the form of Li_2S/Li_2S_2 in a fully discharged state. The physical form of the sulfur compounds is converted from one solid phase to another solid phase, via the intermediate liquid phase, in a discharge process. The same phase transformation order occurs during the charging cycle. These phase transitions are a serious difficulty when producing positive electrodes:

(1) The positive electrode must retain its structural integrity when the solid forms of sulfur turn into liquid. Ideally, when the sulfur forms become liquid in the electrolyte, the positive electrode needs to take a porous structure, of course without breaking down.

(2) In the case where liquid forms are converted back to solid sulfur forms, it is expected that solids will uniformly fill the spaces in the electrode.

(3) Solid forms of sulfur must maintain close electrical contact with the carbon in the electrode.

(4) The electrode must maintain a double continuous porous structure to provide a lithium ion transport path. During cyclic work, highly mobile polysulfide forms cannot be restored to their predicted physical sites.

Therefore, structural damage to the positive electrode during battery operation often causes the failure of Li-S batteries. In order to achieve good ability to operate cyclic Li-S batteries and enable their further development, many materials have been developed for positive electrodes, including carbon-sulfur and polymer-sulfur composites, organic sulfides, inorganic additives, and in parallel also many new binders.

A lot of strategies were used to improve sulfur utilization and reduce the dissolution of polysulfide intermediates. They were, namely: to use sulfur host materials [2–6]; to use protective coating layers [7–11]; and to use an interlayer between cathode and separator [12, 13], promoting the springing up of the sulfur cathodes with high-specific capacities. Most of the reported Li-S batteries

Next-Generation Batteries With Sulfur Cathodes
https://doi.org/10.1016/B978-0-12-816392-4.00005-0

were designed with low-sulfur content (<70 wt%) and/or low-sulfur loading (<2 mg cm^{-2}). The corresponding areal capacities (usually < 2 mAh cm^{-2}) were smaller than the commercial LiBs (4 mAh cm^{-2}) [14–16].

Schematics of a traditional cell and one of the new configurations for rechargeable Li-S batteries are shown in Fig. 4.

5.1 CARBON-SULFUR TYPE COMPOSITES

In the earliest cell configuration, the carbon used as an electronic conductor was mixed with sulfur, resulting in a simple carbon-sulfur composite material [18, 19]. Various carbon materials were used, including Super P (SP) [20–24] and acetylene (AB) carbon black [25–27].

A review of the electrochemical applications of carbonaceous materials is presented, among other things, in [28]. Reference [29] describes the development of carbon-sulfur composites and the application for Li-S batteries. The analyzed groups of carbon-sulfur composites were:

- mesoporous carbon-sulfur composites,
- microporous carbon-sulfur composites,
- hierarchical porous carbon-sulfur composites,
- hollow carbon-sulfur composites, carbon black–sulfur composites,
- carbon nanotube/fiber-sulfur composites, graphene sheet-sulfur composites,
- polyacrylonitrile-derived carbon-sulfur composites,
- flexible carbon-sulfur composites.

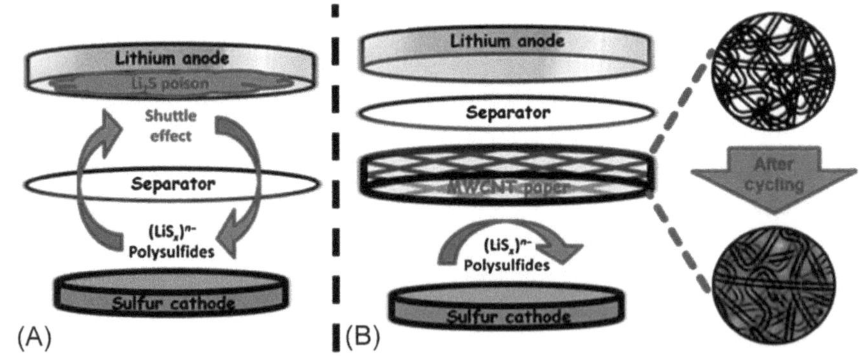

Fig. 4 Schematic cell configuration of rechargeable Li-S batteries: (A) traditional configuration with severe shuttle effect and Li_2S poison problems; (B) new configuration with the MWCNT interlayer. *Taken from M. Liu, X. Qin, Y.-B. He, B. Li, F. Kang, Recent innovative configurations in high-energy lithium–sulfur batteries, J. Mater. Chem. A 5 (2017) 5222–5234.*

Discussions have been devoted to the synthesis approach to the various carbon-sulfur composites, the structural transformation of sulfur, the carbon-sulfur interaction, and the impacts on electrochemical performances.

According to [29], no one carbon form alone can meet the comprehensive performance criteria, but the right combination of these materials can yield the best advantage from the properties of the individual building blocks. Critical factors were suggested for the rational design of advanced carbon-sulfur composites.

The interface between sulfur and carbon is limited due to inhomogeneous mixing [21, 30, 31] and a weak cyclic capacity occurs due to the passivation layer of Li_2S accumulated on the surface of the carbon matrix [21, 22]. The formation of irreversible Li_2S is attributed to structural damage due to the stresses arising during the loading and unloading process [22]. The homogeneity of mixing is improved by heat treatment [32]. Briefly, a mixture of sublimated sulfur and acetylene black AB is milled to obtain homogeneity and heated at 149°C in an argon-filled reactor. The molten sulfur has the lowest viscosity at this temperature and disperses in the pores of AB. The temperature is then increased to 300°C, at which temperature the sulfur is evaporated and it is allowed to diffuse into the nanopores. Acetylene carbonate AB has a specific surface area of 65 $m^2\ g^{-1}$ with an average pore size of about 2.5 nm. The total surface area is reduced to 33.2 $m^2\ g^{-1}$ after the introduction of sulfur through thermal treatment, accompanied by a significant reduction in small pores. These facts indicate that sulfur diffuses into the AB nanopores and disperses very well, as shown in Fig. 5. For comparison, when the acetylene glycol AB composite and sulfur are prepared by ball milling, sulfur covers only the AB surface. The surface of the material prepared in this way decreases from 65 to 8.9 $m^2\ g^{-1}$. The composite prepared by heat treatment provides a large initial discharging capacity of 935 $mAh\,g^{-1}$, and this is maintained at 500 $mAh\,g^{-1}$ after 50 cycles. The improvement of the initial discharging capacity and cyclicity is caused by the trapping of polysulfides in the nanopores of acetylene black AB [32].

Carbon blacks such as SP and AB have relatively small surface areas of 60–70 $m^2\ g^{-1}$, which limits sulfur dispersion. Other carbonaceous materials with large surface areas and small pore sizes, such as activated carbon (AC) [31, 33, 34], Ketjen Black (KB) [30, 35] soot, and carbon paste (CP) [35], are used as receiving materials for the dispersion of sulfur nanoparticles. A high initial discharging capacity of 1180 $mAh\,g^{-1}$, using sulfur in about 70%, is obtained when AC is used as an additional conductive agent [34]. Capacity after 60 cycles shows 60% of the initial discharging capacity. The close contact of sulfur with highly porous AC provides good conductivity to

Fig. 5 SEM images of sulfur (A), acetylene black (AB) (B), sulfur–AB composite prepared by thermal treatment (C), and sulfur–AB composite prepared by ball-milling (D). *Taken from B. Zhang, C. Lai, Z. Zhou, X.P. Gao, Preparation and electrochemical properties of sulfur–acetylene black composites as cathode materials, Electrochim. Acta 54 (2009) 3708–3713.*

composite materials, thus giving excellent electrochemical properties to the composite material. In comparison to the ball milling method, the use of sulfur is improved by heat treatment of the mixture of sulfur and AC [31]. However, worse results are obtained with an increase in sulfur content. When KB mixes with sulfur, a limited improvement in pot life is obtained. Nevertheless, the viability is significantly improved along with the better uniformity of the carbon distribution around the sulfur particles when dry constituents such as sulfur, carbon and binders are ground in a ball mill before preparation of the slurry [30]. Studies on carbon blacks with different adsorption possibilities proved that the key parameter for coal, conditioning good results, is its strong adsorption capacity in relation to polysulfides. A large surface area is less important than the absorption capacity [36].

Reference [12] studied a cell configuration using microporous carbon film as an interlayer covered on the surface of a bare sulfur cathode, efficiently capturing polysulfides and blocking their diffusion to the anode side.

To improve the integrity of the structure, multiwall carbon nanotubes (MWCNTs) and carbon nanofibers (CNFs) are added in addition to AB as additional electronic conductors for the sulfur electrode. The addition of MWCNTs creates a 3-D network, which is more efficient than AB for electronic conduction, and thus improves the initial discharging capacity and capacity maintenance [37]. However, the initial capacity is relatively low due to the inhomogeneous dispersion of sulfur. Sulfur dispersion improves deposition from the gas phase to MWCNT [38]. The initial discharging capacity is about 700 mAh g^{-1}, which is much higher than that observed for the case without the addition of MWCNTs. MWCNTs play important roles in the positive electrode: supplying the nanoreactor to the electrochemical reaction, preventing the dissolution of polysulfides, and tolerating a thick passivation layer of Li_2S. Alternatively, the MWCNT-sulfur nanocomposite is obtained by solvent exchange based on different solubilities of sulfur in various solvents [39]. Modification of MWCNT improves the interaction of MWCNT with solvent and dispersion. In addition, the hydroxyl, carboxyl, and carbonyl groups introduced act as growth points when the sulfur redeposits in the loading cycle [39]. Materials prepared in this way provide initial discharging capacities of 1380 and 1020 mAh g^{-1} after 30 cycles. The CNFs were studied as electronic and physical conductors of the sulfur electrode binder [40]. The improvement of cyclic working capacity of CNF and sulfur composite is attributed to a reduced agglomeration of sulfur or lithium sulfide. Sulfur coating on the surface of various carbon materials improves sulfur dispersion in composite materials. S coated MWCNTs are prepared by maintaining capillarity between liquid sulfur and MWCNTs. With a low surface tension of 61 mN m^{-1}, liquid sulfur wets and fills the walls of carbon nanotubes through capillarity (see Fig. 6) [41, 42].

MWCNTs coated with S show a reversible capacity of 670 mAh g^{-1} after 60 cycles. The cyclicity is much better compared to S-coated carbon black and a simple mixture of S and MWCNT, the former showing a better cyclical ability than the latter [41]. Alternatively, sulfur may be deposited on the carbon surface by a chemical reaction of a solution of sulfide and sulfur dioxide [43]:

$$SO_2 + 2S^{2-} = 3S\downarrow + 2H_2O \tag{12}$$

Sulfur deposition begins on the surface of the carbon and then wraps around the whole carbon particles, creating a core-shell structure. The uniform coating, a few nanometers thick, is formed on the surface of the carbon particles (see Fig. 7). The continuous AC matrix provides electron transport and its surface offers reaction sites. The initial discharging capacity is around 1232 Ah g^{-1}, and this drops to 800 mAh g^{-1} after 50 cycles and remains

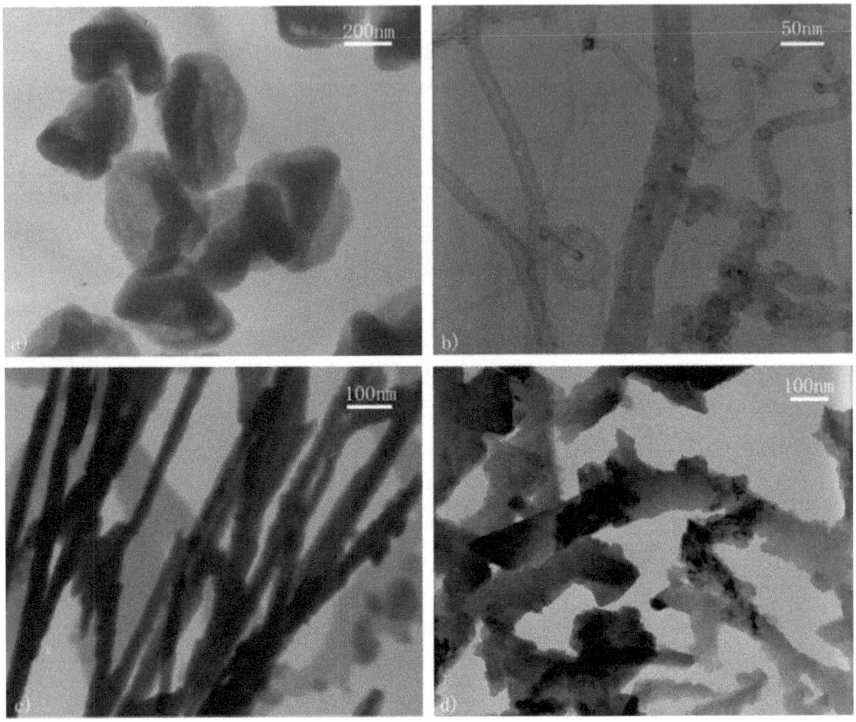

Fig. 6 Images of TEM (A) of elemental sulfur, (B) MWCNTs, (C) S-coated MWCNTs (50% sulfur), and (D) S-coated MWCNTs (80% sulfur). *Taken from L.X. Yuan, H.P. Yuan, X. P. Qiu, L.Q. Chen, W.T. Zhu, Improvement of cycle property of sulfur-coated multi-walled carbon nanotubes composite cathode for lithium/sulfur batteries, J. Power Sources 189 (2009) 1141–1146.*

in this figure. The ability to withstand high currents is improved due to the good electronic conductivity of the composite material.

As reported in [44], binder free vertical aligned (VA) CNT/sulfur composite electrodes with high sulfur loadings up to 70 wt% were synthesized, delivering discharge capacities higher than 800 mAh g^{-1} of the total composite electrode mass.

An improvement of cyclicity is also observed in the sulfur-carbon composite produced by spraying [45]. The porosity of carbon plays an important role in the use of active materials and maintaining capacity during cyclic battery operation. Recent achievements in the synthesis of porous carbon materials with well-defined nanostructures open the possibility of shaping the properties of sulfur-carbon composites. Mesoporous carbon (MPC) with a disordered porous structure was synthesized using the copolymerization

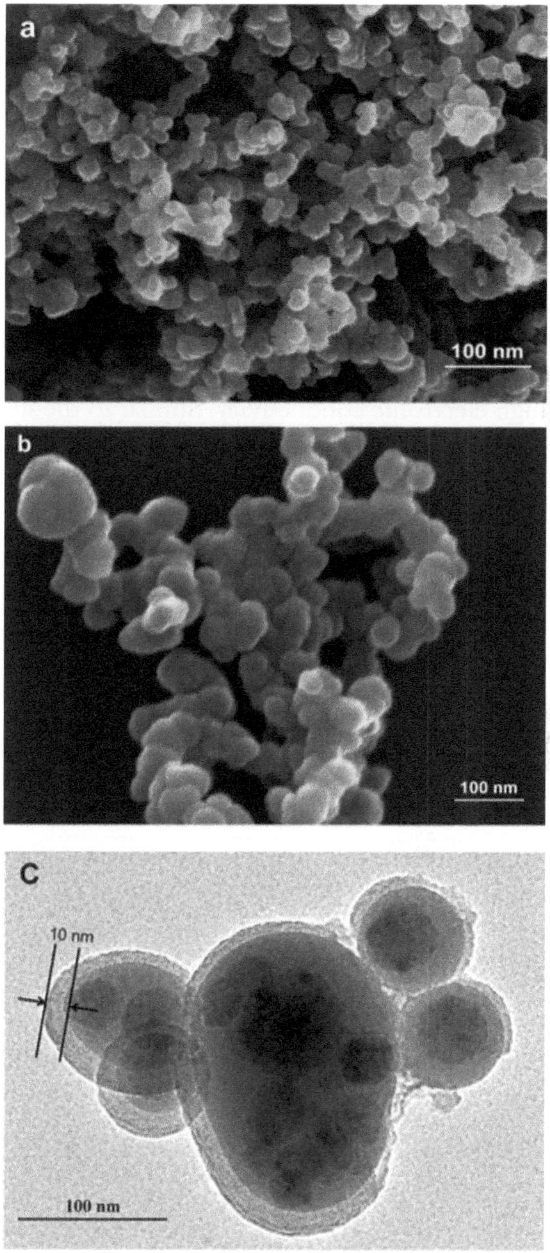

Fig. 7 SEM morphology for super P (A), sulfur-coated carbon (B), and TEM image of sulfur-coated carbon (C). *Taken from C. Wang, J.J. Chen, Y.N. Shi, M.S. Zheng, Q.F. Dong, Preparation and performance of a core–shell carbon/sulfur material for lithium/sulfur battery, Electrochim. Acta 55 (2010) 7010–7015.*

of sodium silicate and sucrose, followed by the removal of the silicate template after the carbonization of organic compounds [46]. The sulfur-mesoporous carbon composite was obtained by coating MPC with sulfur by infiltration in the gas phase [46]. In combination with a liquid ionic electrolyte, the initial discharging capacity and cyclic capacity are better compared to a sulfur electrode prepared by mechanical mixing of soot and sulfur. The improvement is attributed to the strong adsorption of polysulfides to MPC, thus reducing their dissolution in liquid electrolytes. The efficiency of the carbon-sulfur composite material can be further improved by using a highly ordered MPC as the receiving material. CMK-3, ordered MPC material, has the same pore diameter, large pore volume, and interconnected porous structure with high electronic conductivity. Similar to the short rod system, CMK-3 morphology allows access to mesoporous canals. CMK-3 composite and sulfur are produced using the diffusion alloy method. When heated at 155°C, the molten sulfur is impregnated into the channels by capillary forces. Sulfur nanofibers arise inside the porous carbon, forming close contact with the conductive carbon walls (see Fig. 8). The carbon frame structure provides good electrical contact with sulfur and fixes the sulfur in its channels, which helps to stop intermediate polysulfides and facilitate full reduction of sulfur to Li_2S_2/Li_2S (or oxidation to S_8). The retention of various types of polysulfides is enhanced by polymer modification of the carbon surface. The obtained chemical gradient prevents polysulfide diffusion to the electrolyte, and thus the transfer phenomenon is significantly inhibited. The initial capacity is about 1320 mAh g^{-1} and the reversible capacity of 1100 mAh g^{-1} is

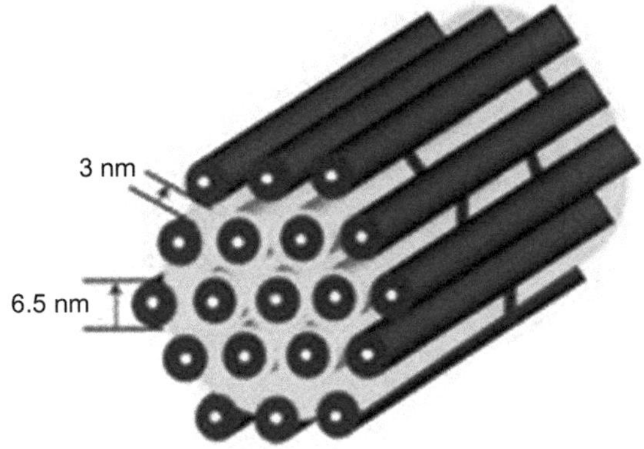

Fig. 8 Sulfur *(yellow)* diagram embedded in the interconnected pore structure of mesoporous coal, CMK-3. *Taken from X.L. Ji, K.T. Lee, L.F. Nazar, A highly ordered nanostructured carbon–sulphur cathode for lithium–sulphur batteries, Nat. Mater. 8 (2009) 500–506.*

maintained after 20 cycles. The ability to work cyclically is improved due to better control of morphology, in addition to polymeric modification [2].

Highly porous carbon (HPC) can contain 57% sulfur while maintaining a porous structure. After completing the introduction of sulfur into the HPC micropores, its surface area decreased from 1473 to 24 $m^2\,g^{-1}$. This composite provides an initial capacity of 1155 $mAh\,g^{-1}$ and a reversible capacity of 745 $mAh\,g^{-1}$ after 84 cycles. The ability to withstand the maximum currents is better, although it is paid for by the low capacity. The ultrafine porous structure delays dissolution of polysulfides in electrolytes due to the strong adsorption of polysulfides in microporous carbon [47]. However, increasing the sulfur content to 75% results in the disintegration of the macroporous structure, which leads to lower loading/unloading capacity. The hierarchically organized sulfur-carbon composite was noted as a positive electrode for the lithium-sulfur battery [48]. The MPC, which was prepared by a soft template–enhanced synthesis [49–52], was activated using potassium hydroxide. The resulting activated MPC has a bimodal porous structure with two pore size distributions, with dimensions of 2

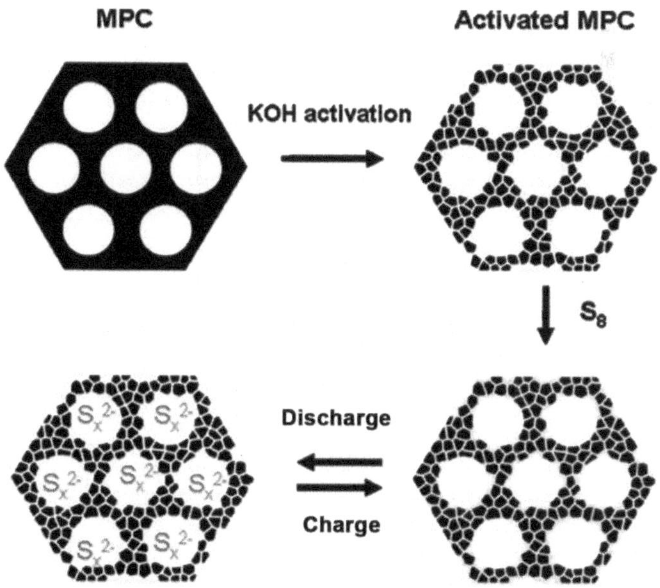

Fig. 9 Illustration of the S/C cathode composite material using bimodal porous coal as a support. *Taken from C.D. Liang, N.J. Dudney, J.Y. Howe, Hierarchically structured sulfur/carbon nanocomposite material for high-energy lithium battery, Chem. Mater. 21 (2009) 4724–4730.*

and 7.3 nm, as shown in Fig. 9. The micropores exist as intrawall porosity in the mesopores. The micropores function as microreservoirs for elemental sulfur, the large surface area providing significant contact with insulating sulfur and high electrical conductivity. Mesopores retain electrolyte, facilitate lithium-ion transport, and limit polysulfides. The MPC-S composite is made using the liquid infiltration method. When the sulfur content is 11.7%, the initial discharging capacity is 1585 mAh g^{-1}, which corresponds to 95% utilization of sulfur with respect to its theoretical capacity of 1675 mAh g^{-1}. The original MPC with 25.2% sulfur provides an initial discharging capacity of 1136 mAh g^{-1}. However, WVA-1500 (which mainly contains micropores) with 25.2% sulfur has a very low initial capacity of 388 mAh g^{-1}. Therefore mesopores are responsible for high initial capacitance, while micropores contribute to maintaining the capacity and high ability to withstand the maximum currents (see Fig. 10). To increase long-term stability, the sulfur-carbon composite of the sphere is prepared by placing sulfur in the micropores of carbon spheres by thermal treatment of MPC [49].

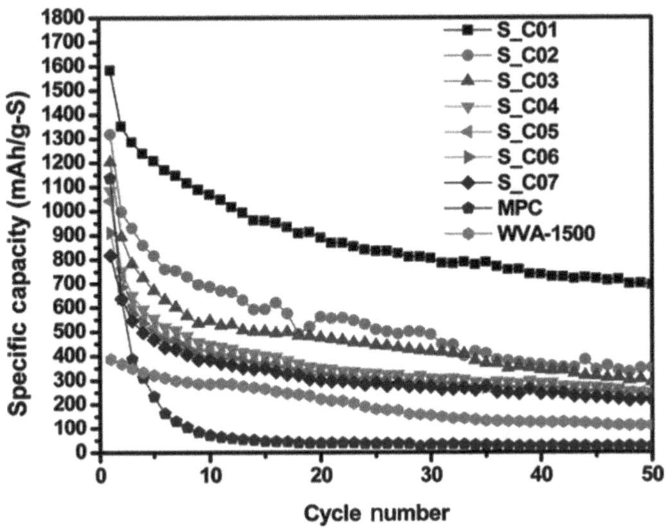

Fig. 10 Specific capacity of discharging active MPC/sulfur composite. The sulfur content in the samples varied from 12% wt for S_C01, through 24% wt for S_C03, to 52% wt for S_C07. A sample of mesoporous carbon (MPC) was synthesized by the soft template method. After activation of KOH, the elemental sulfur was loaded into activated mesoporous carbon (α-MPC) by infiltration in a liquid CS2 solution containing 10% by weight of sulfur. The WVA-1500 sample was made from activated microporous WVA-1500 carbon produced by Mead Westvaco Corp. *Taken from C.D. Liang, N.J. Dudney, J.Y. Howe, Hierarchically structured sulfur/carbon nanocomposite material for high-energy lithium battery, Chem. Mater. 21 (2009) 4724–4730.*

The carbon beads obtained in the synthesis have a narrow micropore size distribution of approximately 0.7 nm and a large specific surface area of $844\,m^2\,g^{-1}$ (Fig. 11). The carbon spheres can accommodate several annular S_8 crowns and/or short elemental sulfur chains. The total contact of sulfur and carbon is significantly improved. The total area is reduced to $6.5\,m^2\,g^{-1}$ after filling the carbon balls with 42% sulfur by weight. The volume of micropores is almost completely filled with elemental sulfur, which ensures high sulfur dispersion in the micropores of carbon spheres. The high initial discharging capacity of $1333\,mAh\,g^{-1}$ and the reversible capacity of $1006\,mAh\,g^{-1}$ are obtained at a current of $40\,mA\,g^{-1}$. Interestingly, the high reversible capacity of $730\,mAh\,g^{-1}$ is maintained at a high current of $1200\,mA\,g^{-1}$. The narrow distribution of micropores (0.7 nm) rather than area and total pore volume is important for controlling the adsorption capacity of sulfur-containing compounds [50]. In addition to good electrical conductivity, microporous carbon beads reduce elemental sulfur and subsequent polysulfides during cyclic work, due to its strong adsorption, minimizing the

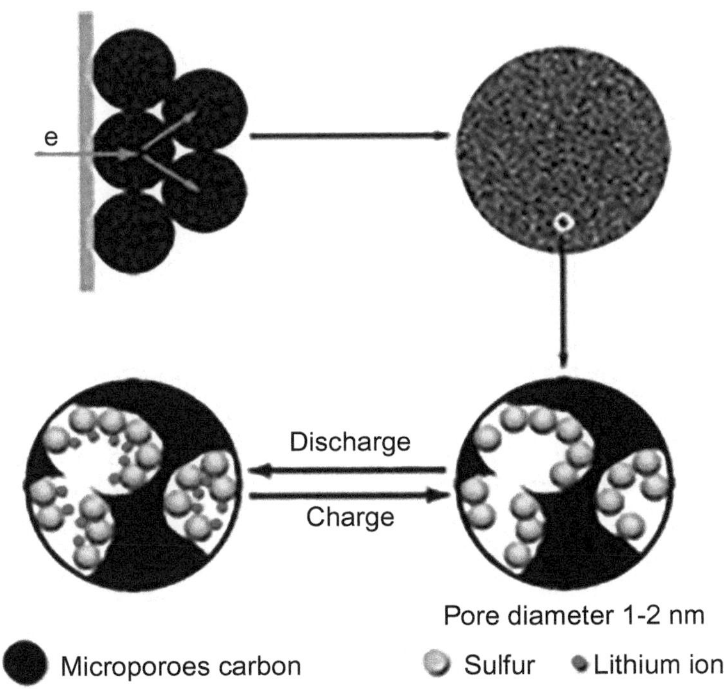

Fig. 11 Diagram of a limited electrochemical reaction process inside micropores of sulfur-carbon composite spherical particles. *Taken from B. Zhang, X. Qin, G.R. Li, X.P. Gao, Enhancement of long stability of sulfur cathode by encapsulating sulfur into micropores of carbon spheres, Energy Environ. Sci. 3 (2010) 1531–1537.*

transfer phenomenon, which causes the loss of active substances and formation of a thick Li_2S insulating layer on the surface of composite electrodes [49]. Therefore the ability to work cyclically and the ability to withstand the maximum currents are much better.

The S cathode performance can be improved using mesoporous hollow carbon spheres, as reported also in [53, 54].

As explained in [55], encapsulation of nanostructured S and attenuation of polysulfide dissolution using melting and infusion of S into other materials matrix have limitations in realizing the currently needed S cathode. At the macroscopic scale, electrode homogeneity and integrity, ensuring cathode performance, were very difficult to obtain for the top-down approaches of melting and infusing S into the matrix. In many cases, S precipitation on the outer surface of the host matrix could not be avoided. The additional processes (carbon disulfide washing or heating) were needed to remove the inhomogeneously deposited S to avoid rapid capacity decay. This problem has also led to great variation in battery performance, even with similar electrode structure and composition.

The use of metal oxides as adsorbent materials is disclosed in the last patent [51]. However, these materials are not effective in improving the viability due to their small area and insulating character. Nanodimensional $Mg_{0.6}Ni_{0.4}O$ showed adsorptive properties and the effect of the catalyst, while the addition of this material to the sulfur-carbon composite electrode increases the initial capacity and cyclic operation due to the reduction of polysulfide dissolution [52]. Improving the ability to withstand the maximum currents, however, is attributed to the catalytic effect of the redox type Li-S reaction. Similar results are obtained when the nanosized $Mg_{0.8}Cu_{0.2}O$ is added to composite electrodes V_2O_5-S [56]. Dissolution of polysulfides is also reduced by the addition of nanosized Al_2O_3 in the sulfur-carbon composite electrode, thus improving the initial capacity and the ability to work cyclically [57]. The use of the prelit C-S composite as a positive electrode [23, 58, 59] enables the use of nonlithium metals as negative electrodes and avoidance of problems with the safety of the metallic lithium anode. In the work described in [58], an attempt was made to use an electrochemically compressed C-S composite electrode as a cathode, but enlargement is a challenge [60]. A practical approach to the preparation of prelit C-S composite is ball milling of pure crystalline lithium sulfide and SP coal under an argon atmosphere [23]. In combination with a Sn-C negative electrode and a polymeric electrolyte (Fig. 12), the energy density of 1000 or 2000 $Wh\,kg^{-1}$ is provided, depending on whether the weight of the active

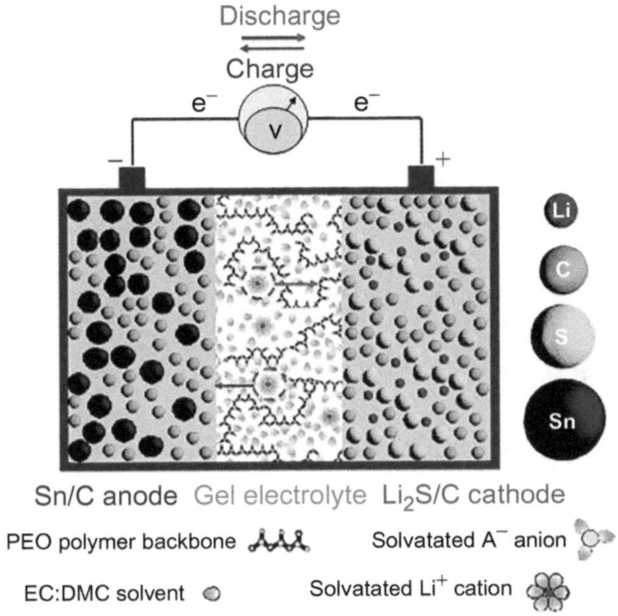

Fig. 12 Scheme of Sn/C/CGPE/Li_2S/C polymer battery. *Taken from J. Hassoun, B. Scrosati, A high-performance polymer tin sulfur lithium ion battery, Angew. Chem. Int. Ed. 49 (2010) 2371–2374.*

substance Li_2S-C or Li_2S is taken into account. Alternatively, the Li_2S-CMK-3 MPC composite was prepared by spraying the mesopores with sulfur and then converting the trapped sulfur to lithium sulfide by reaction with n-butyl lithium [59].

When connected to an anode from a silicon nanowire (Fig. 13), the initial discharging capacity is 573 mAh g^{-1}, stabilizing after five cycles. A small pore size with a strong limiting effect is essential for achieving good cyclic work efficiency. Yet another ordered mesoporous carbon was synthesized using SBA-15 silica as a hard template and humic acid as a carbon source [61]. The obtained carbon material was characterized by a hexagonal order of pores with an average diameter of 7.6 nm and a large specific surface area of 670 m^2 g^{-1}. Such synthesized carbon served as a component of the sulfur cathode in the Li-S cell. At 20% of the content in the composite cathode and without any auxiliary conductive additives, the material was characterized by a much more favorable capacity relative to conventional carbon black and commercial CMK-3 mesoporous coal (up to 1200 mAh g^{-1}), as well as improved cyclic stability and transferability rated currents. This was attributed to a favorable morphology and a porous structure of carbon obtained from

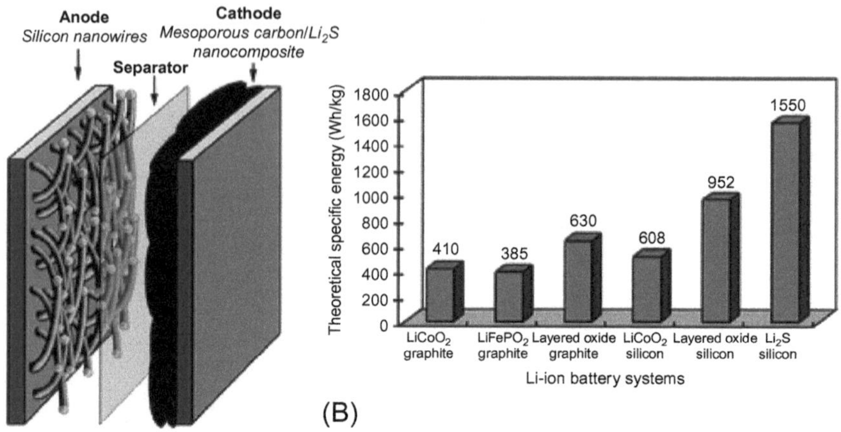

Fig. 13 Diagram of Li_2S/Si battery cell structure (A) and comparison of the specific energy of various Li-ion battery systems (B). *Taken from Y. Yang, M.T. McDowell, A. Jackson, J.J. Cha, S.S. Hong, Y. Cui, New nanostructured Li_2S/silicon rechargeable battery with high specific energy, Nano Lett. 10 (2010) 1486–1491.*

humic acid. Another large surface area carbon was obtained from biomass [62]. It was prepared by chemical processing of fruit skins with potassium carbonate. This carbon was tested as a component of the Li-S sulfuric cathode battery. The synthesized material was characterized by a hierarchical micro-macroporous structure with a system of connected microfibers similar to a sponge. The composite sulfur cathode produced using this carbon showed a very large initial discharge capacity of 886 $mAh\,g^{-1}$ at a load of 0.1 C, which remained very stable over the next 50 charge/discharge cycles (capacity losses below 1%). The excellent reversibility and stability of the cyclic work of this composite cathode has been attributed to the unique combination of micro- and macroporosity, as well as chemical surface interactions that allow the intermediate polysulfides to be retained inside the carbon structure without excessive dissolution in the electrolyte region.

According to [17], reduced graphene oxide (rGO), carbon nanofibers (CNFs), self-assembled carbon nanotubes (CNTs), self-assembly polypyrrole nanotubes, and carbonized bacterial cellulose can be interlayers for Li-S cells due to their conductive framework for electron conduction and porous structure for ion transfer, and adsorption of sulfur species.

Paper [63] presents nanocomposite cathode materials consisting of sulfur (80% by weight), embedded in three-dimensional, doped graphene oxide (N-3D-rGO), through a controlled method of sulfate impregnation. The addition of nitrogen increases the surface area for prismatic graphene 10 times and the pore volume 7 times. These design features allow the cathode to keep more sulfur. It adsorbs polysulfides and prevents them from being detached from the matrix material, thus ensuring stable results of cyclic work. The saturation method of the sulfur solution allows obtaining a uniform distribution of nanosulfur in a controlled manner. The resulting material provides a high initial discharging capacity of 1042 and 916 mAh g^{-1} with good holding capacity at 94.8% and 81.9% respectively at 0.2°C and 0.5°C after 100 cycles.

Reference [64] reported a hollow carbon nanofiber–encapsulated sulfur cathode for effective trapping of polysulfides (Fig. 14). The hollow carbon nanofiber arrays were fabricated using anodic aluminum oxide (AAO) templates, through thermal carbonization of polystyrene. The AAO template facilitated sulfur infusion into the hollow fibers and prevented sulfur from coating onto the exterior carbon wall. The high aspect ratio of the carbon nanofibers provided an ideal structure for trapping polysulfides, and the thin carbon wall allowed rapid transport of lithium ions. The small dimension of such nanofibers provided a large surface area per unit mass for Li_2S deposition during cycling and reduced pulverization of electrode materials due to volumetric expansion. A high specific capacity of about 730 mAh g^{-1} occurred at C/5 rate after 150 cycles of charge/discharge. The introduction of $LiNO_3$ additive to the electrolyte improved the coulombic efficiency to over 99% at C/5.

Improvement in the S cathode performance was obtained using hollow/porous carbon nanofibers, as reported in [3, 64, 65], or activated carbon fiber, as shown in [66].

The heteroatom doped CNT (one-dimensional allotropic carbon variety) improves the overall conductivity of the electrodes and overcomes limitations associated with lithium polysulfides [67–70]. Nitrogen doping shows

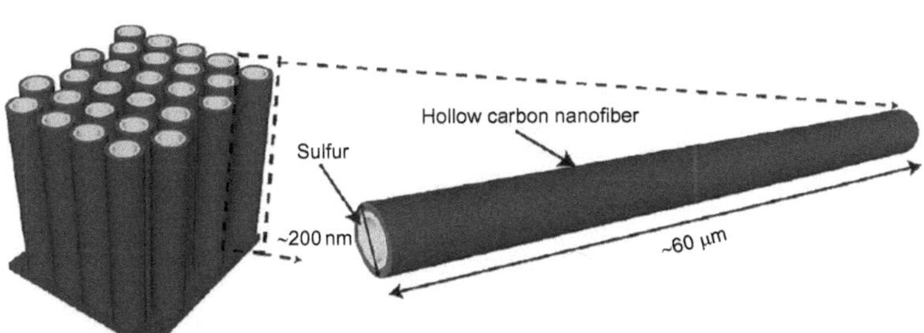

Fig. 14 The design principle showing the high aspect ratio of the hollow carbon nanofiber for effective trapping of polysulfides. *Taken from G. Zheng, Y. Yang, J.J. Cha, S.S Hong, Y. Cui, Hollow carbon nanofiber-encapsulated sulfur cathodes for high specific capacity rechargeable lithium batteries, Nano Lett 11(10) (2011) 4462–4467.*

significant chemisorption of lithium polysulfides due to the strong interaction of Li-N, which is associated with the property of nitrogen donor electrons [67]. Lithium-sulfur batteries with a long and stable service life under cyclic conditions were obtained by using CNTs doped with nitrogen or graphene as matrices [67, 68, 71–74]. In spite of this, the relatively low specific surface area of CNT limits the sulfur and polysulfide deposition, which means less use of sulfur and reduced energy density [75–77]. CNTs tend to aggregate, thanks to the strong interactions of van der Waals forces, which reduce their physical properties [78]. In some research cases, the nitrogen used for doping came from external sources such as ammonia [79], urea [80], pyridine [81], or melamine [82]. Wang et al. [80] used urea as a source of nitrogen to synthesize nitrogen-doped graphene sheets with the hydrothermal method, which allowed increased long-term battery stability to be obtained (578.5 mAh g^{-1} capacity remaining at 1C load within 500 cycles).

Improvement of the S cathode performance can also be obtained using graphene oxides, as shown in [83].

Reference [84] reported the synthesis of a graphene-sulfur composite material by wrapping poly(ethylene glycol) (PEG) coated submicrometer sulfur particles with mildly oxidized graphene oxide sheets decorated by carbon black nanoparticles (Fig. 15). The PEG and graphene coating layers allowed accommodating volume expansion of the coated sulfur particles during discharge, trapping soluble polysulfide intermediates, and rendering the sulfur particles electrically conducting. The resulting graphene-sulfur composite showed high and stable specific capacities up to 600 mAh g^{-1} over more than 100 cycles.

Reference [85] identified a new capacity fading mechanism of the sulfur cathodes, relating to Li_xS detachment from the carbon surface during the discharge process. Such a capacity fading mechanism can be overcome by introducing amphiphilic polymers to modify the carbon surface, rendering strong interactions between the nonpolar carbon and the polar Li_xS clusters (Fig. 16). The modified sulfur cathode showed high cycling performance with specific capacity close to 1180 mAh g^{-1} at a C/5 current rate. Capacity retention of 80% was achieved over 300 cycles at C/2.

According to [29], for cyclic stability of sulfur, micropore adsorption or possible intercalation between graphene layers lead to the best results, while surface functionalization can improve this further. Mesopores, or loose confinement, e.g., graphene wrapping, improve reaction kinetics. The large pore volumes of mesoporous carbon and graphene sheets are also advantageous for high loading of sulfur, which gives high capacity of the C-S

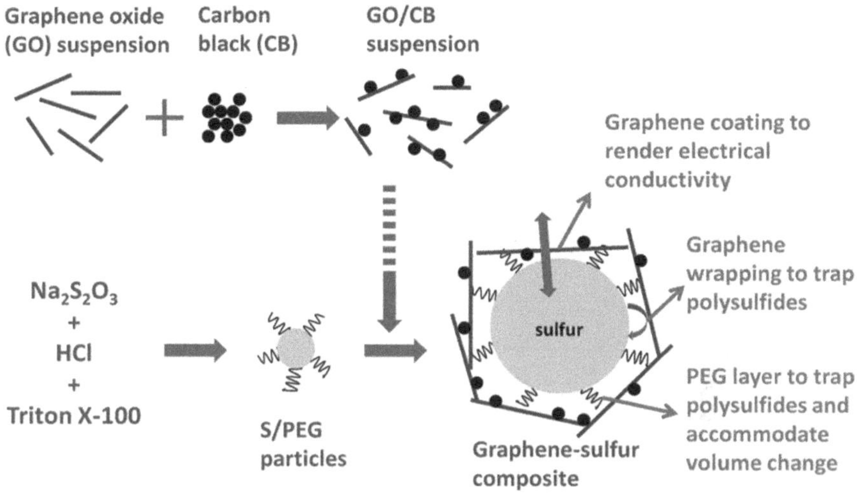

Fig. 15 Schematic of the synthesis steps for a graphene-sulfur composite, with a proposed schematic structure of the composite. *Taken from H. Wang, Y. Yang, Y. Liang, J.T. Robinson, Y. Li, A. Jackson, Y. Cui, H. Dai, Graphene-wrapped sulfur particles as a rechargeable lithium–sulfur battery cathode material with high capacity and cycling stability, Nano Lett. 11 (7) (2011) 2644–2647.*

composite. Partially filled pores can compensate for the volume change of sulfur-lithium sulfides, so the optimal loading of sulfur is a balance between the maximum capacity and the needed allowance for the volume change to ensure stability. Hollow carbons with rigid shells and large internal voids are desirable for this purpose. Graphene sheets with good flexibility are useful for buffering volume change.

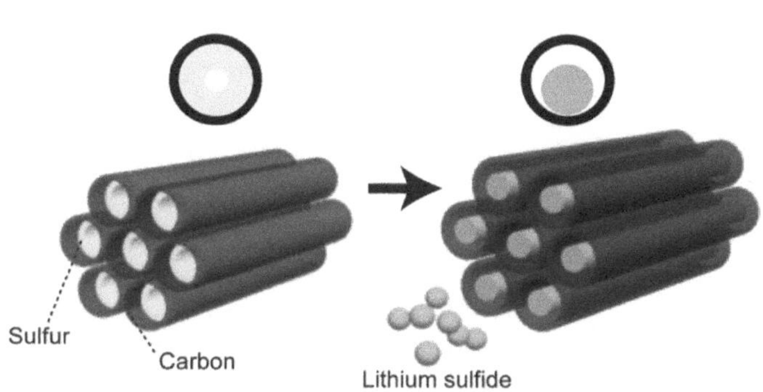

Fig. 16 Discharge profile of the sulfur cathode. Insets are schematics showing the morphological change of sulfur cathode after discharge. *Taken from G. Zheng, Q. Zhang, J.J. Cha, Y. Yang, W. Li, Z.W. Seh, Y. Cui, Amphiphilic surface modification of hollow carbon nanofibers for improved cycle life of lithium sulfur batteries, Nano Lett. 13(3) (2013) 1265–1270.*

During studies described in [86] filamentous fungus was used as a carbonizable binder to fabricate a graphene-embedded carbon fiber film. The N,O-doped film served as a superior conductive interlayer. Therefore, doping with miscellaneous elements enhanced the interaction between the polysulfides and carbon matrix.

Reference [87] analyzed the interactions between lithium polysulfides and N-doped graphene (N-G) with different doping configurations. Only N-G with clustered pyridinic N-dopants could effectively attract and tightly anchor soluble polysulfides by means of their large binding energies, which was due to the enhanced attraction between Li ions in polysulfides and pyridinic N-dopants, and the additional attraction between the sulfuric anions in polysulfides and Li ions captured by pyridinic N-dopants.

According to [17], the relatively weak physical adsorption between nonpolar carbon and polar polysulfides still limits their capture ability and recycling efficiency.

Reference [88] showed an interlayer comprised of an outer cyclized polyacrylonitrile (PAN) network and inner carbon nanofiber skeleton (CP@CNF) based on a simple dip-coating and thermal cyclization treatment. The good electrochemical performance resulted from the strong interaction between the polysulfides and outer nitrogen-doped groups; particularly, the pyridinic nitrogen restricted diffusion and increased the reutilization of polysulfides.

In Reference [89], an ultrathin yet multifunctional polysulfide blocking layer composed of active carbon nanoparticles and conductive polymer (PEDOT:PSS) was reported, which was coated on the cathode by electrospray to restrain the shuttle effect of Li-S cells. The thin interlayer allowed fast ion diffusion but still retained a high efficiency of blocking polysulfides from a combination of physical and chemical confinements.

In studies reported in [9], a conducting polymer was applied to minimize the diffusion of polysulfides out of the mesoporous carbon matrix by coating poly(3,4-ethylenedioxythiophene)-poly(styrene sulfonate) (PEDOT:PSS) onto mesoporous carbon/sulfur particles (Fig. 17). After surface coating, the coulomb efficiency of the sulfur electrode was improved from 93% to 97%, and capacity decay was reduced from 40%/100 cycles to 15%/100 cycles. Moreover, the discharge capacity with the polymer coating was 10% higher than the bare counterpart, with an initial discharge capacity of 1140 $mAh\,g^{-1}$ and a stable discharge capacity of >600 $mAh\,g^{-1}$ after 150 cycles at C/5 rate.

According to [17], for inorganic decorated carbon-based interlayers, TiO_2/graphene, TiO_2/CNF, ZnO/CNF were introduced into the system both as

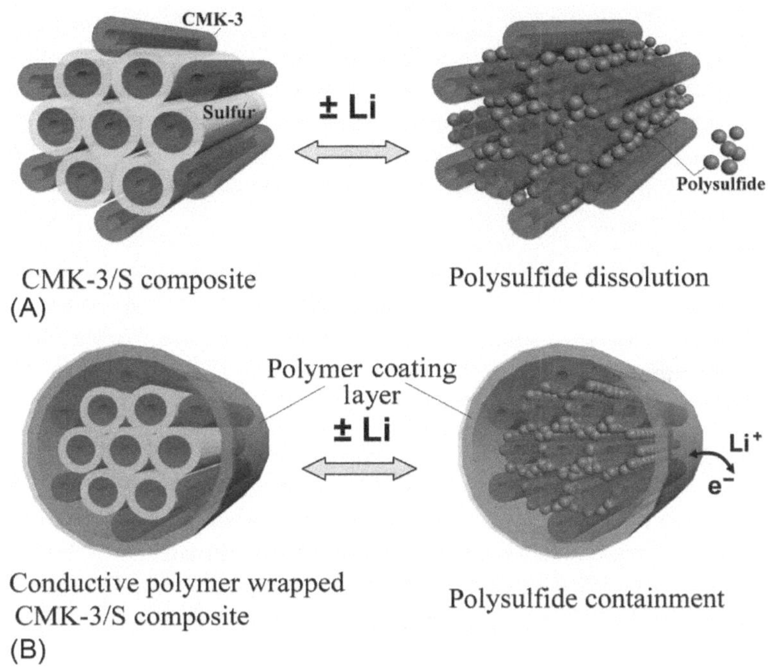

Fig. 17 Scheme of PEDOT:PSS-coated CMK-3/sulfur composite for improving the cathode performance: (A) In bare CMK-3/S particles, polysulfides still diffuse out of the carbon matrix during lithiation/delithiation. (B) With conductive polymer coating layer, polysulfides could be confined within the carbon matrix. Lithium ions and electrons can move through this polymer layer. *Taken from Y. Yang, G. Yu, J.J. Cha, H. Wu, M. Vosgueritchian, Y. Yao, Z. Bao, Y. Cui, Improving the performance of lithium–sulfur batteries by conductive polymer coating, ACS Nano 5(11) (2011) 9187–9193.*

physical barriers (nonpolar) and chemical capturers (polar) to inhibit the shuttling of polysulfides, meanwhile allowing the transport of lithium ions.

An interesting strategy was elaborated in [90] to limit dissolution of intermediate polysulfide reaction species into the electrolyte. It utilized absorption of the intermediate polysulfides by a porous silica embedded within the carbon-sulfur composite that not only absorbed the polysulfides by means of weak binding, but also permitted reversible desorption and release. It functioned as an internal polysulfide reservoir during the reversible electrochemical process to give rise to long-term stabilization and improved coulombic efficiency.

A similar strategy was used during studies described in [91]. The capacity retention and cycle life of the Li-S cell was increased through the use of nanocrystalline and mesoporous titania additives as polysulfide reservoirs.

It also examined the role of surface adsorption vs. pore absorption. It was found that the soluble lithium polysulfides were preferentially absorbed within the pores of the nanoporous titania at intermediate discharge/charge. This provided the major factor in stabilizing capacity although surface binding (adsorption) also played a more minor role. A cell containing TiO_2 with a 5 nm pore diameter exhibited a 37% greater discharge capacity retention after 100 cycles than a cell without the titania additive, which was optimum compared to the other titania that were examined.

Also [92] generalized the use of polysulfide reservoirs to other mesoporous oxide structure, such as metal-organic framework, obtaining 80% capacity retention after 80 cycles.

According to [93], cobalt-embedded nitrogen-doped hollow carbon nanorods (Co@NHCRs) were reported to be employed as sulfur hosts (Fig. 18). Cobalt, as a functional modifier, can strongly adsorb and anchor the Li_2S_2/Li_2S particles, preventing the loss of active mass and maintaining good electrical contact with the conductive carbon matrix.

In Reference [94] it was found that layered structure vanadium pentoxide (V_2O_5) exhibited the strongest chemical interaction with Li_2Sn clusters.

According to [95], V_2O_5 can act as a redox mediator to oxidize polysulfides to thiosulfate/polythionate groups and chemically bond them on the surface of reduced metal oxides due to its higher redox potential than the traditional widely adopted materials, such as TiO_2 [96].

As reported in [97], a V_2O_5 decorated carbon nanofiber (VCNF) membrane was synthesized and utilized as an interlayer in Li-S batteries.

According to [17], the well-decorated V_2O_5 component not only anchors polysulfides through strong interactions, but also effectively suppresses the cell's self-discharge behavior due to its voltage regulation function.

Reference [98] presented a three-dimensional hierarchical carbon material (MCCNT), composed of mesoporous carbon in spherical cores and

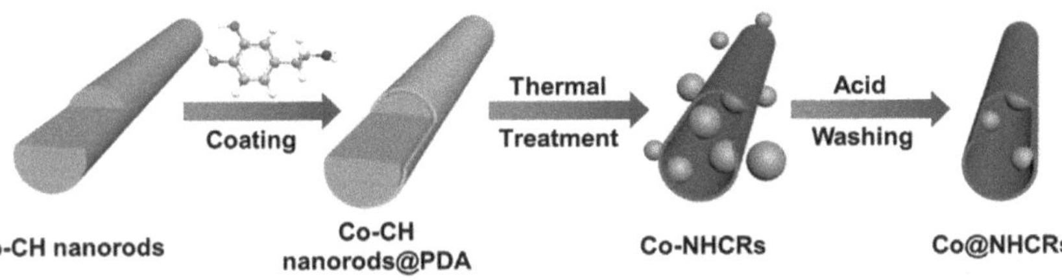

Fig. 18 A schematic illustration for the fabrication of Co@NHCRs. *Taken from M. Zhang, C. Yu, C. Zhao, X. Song, X. Han, S. Liu, C. Hao, J. Qiu, Cobalt-embedded nitrogen-doped hollow carbon nanorods for synergistically immobilizing the discharge products in lithium–sulfur battery, Energy Storage Mater. 5 (2016) 223–229.*

nitrogen-rich CNT coatings, obtained in aerosol sprays and subsequent chemical vapor deposition (CVD). Due to their well-defined porous structure and advantageous conductive structure, MCCNTs are used as a potential matrix for sulfur in Li-S batteries, produced using the classical alloy-diffusion method. During cyclic loading with a current density of 0.2 C (1 C = 1675 mAh g^{-1}), the battery achieved an initial capacity of 1448.7 mAh g^{-1}. Even if the current density increased to 1 C, after 300 cycles, the specific capacity remained at 534.6 mAh g^{-1}. Increased electrochemical efficiency may have resulted from the hybrid structure of the MCCNT, in which the porous core acts as a matrix that maintains sulfur and accommodates volume expansion, while external CNTs provide excellent conduction conditions for electrons and ions. In addition, nitrogen doped in situ on the surface of the CNT enables efficient capture of lithium polysulfides, which leads to a significant improvement in cyclic battery performance. During the investigations described in [99], nanoparticles of carbon from novolak were used as a conductive cathode matrix in the lithium-sulfur battery. Self-emulsifying synthesis was used to obtain submicron carbon beads derived from novolak and containing nanopores. Already after pyrolysis, the carbon balls showed a specific surface area of 640 m^2 g^{-1}, which after physical activation increased to 2080 m^2 g^{-1}. The inactivated and activated carbon balls represented nanoporous carbon with a sufficiently medium and large surface area, which allowed assessment of the influence of porosity on the electrochemical properties of cathodes of Li-S batteries. Carbon/sulfur hybrids were obtained by two different methods of sulfur penetration: infusion of molten sulfur (annealing) and generation of sulfur *in situ* from sodium thiosulphate. For activated carbon beads and the use of molten sulfur infusion, the best performance was obtained (~880 mAh gS^{-1} at low current load in the 5th cycle) and high stability of performance (>600 mAh gS^{-1} after 100 cycles).

According to [17], to increase the areal mass loading and utilization of sulfur, gradient-structured cathodes were developed.

Reference [100] proposed an all graphene structure for the sulfur cathode with highly conductive graphene as the current collector and partially oxygenated graphene as a polysulfide adsorption layer, in which the highly porous graphene with a high pore volume of 3.51 cm^3 g^{-1} was the sulfur host, enabling a high sulfur content of 80 wt%. This cathode sulfur loading was of 5 mg cm^{-2}, allowing both a high initial gravimetric specific capacity (1500 mA h g^{-1}) and areal specific capacity (7.5 mA h cm^{-2}).

As reported in [101], a hollow carbon fiber foam (HCFF) was used as a 3D conductive scaffold-like current collector to accommodate a large amount of sulfur–multiwall carbon nanotube (MWCNT)–carbon black (CB) hybrids. The sulfur loading could be increased to 21.2 mg cm^{-2} with the assistance of a thin HCFF interlayer, and a high capacity retention rate of 70% over 150 cycles was achieved.

In [102] a layer-by-layer strategy was described for high areal-capacity sulfur cathodes by directly splinting commercial sulfur powder between porous carbon nanofiber layers. Such a layer-by-layer cathode allowed fast ion and electron transport and maintained the soluble polysulfide intermediates within the electrode. Even with a six-layer cathode, in which the areal sulfur loading was of 11.4 mg cm^{-2}, an initial discharge capacity of 995 mA h g^{-1}, corresponding to a high areal capacity of 11.3 mA h cm^{-2}, could be achieved.

Reference [103] presented a bilayer design composed of a lower pure sulfur and upper carbon current collector. The pure sulfur electrode showed low polarization, high sulfur utilization, and cycle stability even with an ultrahigh sulfur loading of 13.9 mg cm^{-2}.

The use of large-sized sulfur particles only slightly decreases the rate of conversion from S_8 to S_4^{2-} during the initial discharge and has a negligible impact on the subsequent electrochemical performance. The fabrication of bare sulfur electrodes is analogous to that of electrodes for traditional lithium-ion batteries.

According to [104], a core-shell cathode with a pure sulfur core shielded within a conductive shell-shaped electrode was developed. The electrode configuration allowed a high sulfur loading of up to 30 mg cm^{-2} and sulfur content approaching 70 wt%. The core-shell cathodes showed superior dynamic and static electrochemical stability.

As stated in [60], the areal mass loading of active materials strongly affects the production process. The hierarchical gradient cathode is a unique configuration to solve the shuttle problem in Li-S systems. However, for the gradient cathode, due to the unsatisfactory fact of decreased specific energy caused by excess carbon material or liquid organic electrolyte, the areal mass loading of sulfur should be optimized.

5.2 POLYMER-SULFUR TYPE COMPOSITES

In order to prevent sulfur agglomeration, the accumulation of irreversible Li_2S in the cathode, and the dissolution of polysulfides in the electrolyte, a molecularly mixed polymer-sulfur composite was developed [105]. The polymer-sulfur composite was prepared by sulfur intercalation to conductive polymers, such as polyacrylonitrile (PAN), according to Eq. (13) [20, 31, 60, 105, 106]. The initial discharging capacity of the composite is 850 mAh g^{-1} based on total mass, which means that almost all sulfur atoms have been reduced to Li_2S. The reversible capacity remains above 600 mAh g^{-1} after 50 cycles and almost no self-discharge is observed when fully charged after one month on the shelf. Mixing at the molecular level of PAS and S inhibits the dissolution of polysulfides in electrolytes [105]. The ability to work with the PAN-S cyclic composite is improved by heating the PAN mixture and sulfur at a temperature of 450°C [107]. The reversible capacity of 470 mAh g^{-1} is delivered after 380 cycles. Further studies of PAN-S composites are described in [58, 108–111]. PAN precursors with a narrow molecular weight range and high structural purity resulted in PAN-S composite materials having the highest electrochemical efficiency [112]. The PAN-S composite also has a good discharging capacity of 632–854 mAh g^{-1} between −20°C and 60°C, with higher capacity at higher temperature [108]. Other polymer-sulfur composites based on polypyrrole (PpY) [113–115], polypyrrole-*co*-aniline (PPyA) [116], and polythiophene [117] also showed improvement in cyclic work efficiency.

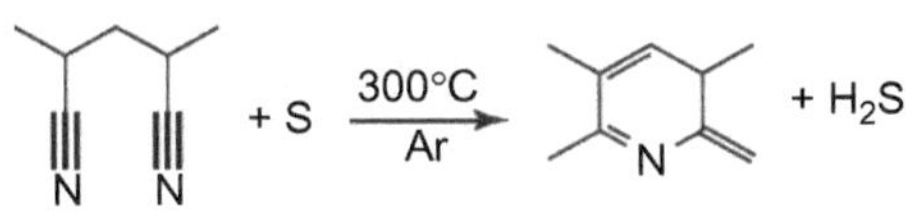

(13)

According to [55], monodisperse polymer (polyvinylpyrrolidone)-encapsulated hollow sulfur nanospheres for sulfur cathode were elaborated, allowing control over electrode design from nanoscale to macroscale. High specific discharge capacities at different current rates (1.179, 1.018, and 990 mAh g^{-1} at C/10, C/5, and C/2, respectively) and excellent capacity retention of 77.6% (at C/5) and 73.4% (at C/2) after 300 and 500 cycles, respectively. Over a long-term cycling of 1000 cycles at C/2, a capacity decay as low as 0.046% per cycle and an average coulombic efficiency of 98.5% was achieved. A simple modification on the sulfur nanosphere surface with a layer of conducting polymer, poly(3,4-ethylenedioxythiophene), allowed the sulfur cathode to achieve excellent high-rate capability, showing a high reversible capacity of 849 and 610 mAh g^{-1} at 2C and 4C, respectively.

Reference [118] studied the influence of different conductive polymers on the sulfur cathode based on conductive polymer-coated hollow sulfur nanospheres with high uniformity. Three conductive polymers, polyaniline

(PANI), polypyrrole (PPY), and poly(3,4-ethylenedioxythiophene) (PEDOT), were coated, respectively, onto monodisperse hollow sulfur nanospheres (Fig. 19) through a polymerization process. It was found that the capability of these three polymers in improving long-term cycling stability and high-rate performance of the sulfur cathode decreased in the order of PEDOT > PPY > PANI. High specific capacities and excellent cycle life were obtained for sulfur cathodes made from these conductive polymer-coated hollow sulfur nanospheres.

5.3 ORGANIC SULFIDES

Organic sulfides were studied as cathode materials based on the redox pair of S-S bonds in organic disulfides or thiolates [119, 120]:

$$\mathrm{RSSR} + 2\mathrm{e} = 2\mathrm{RS}- \text{ (R is an N}-\text{containing alkyl group)} \tag{14}$$

Using RSSR as a cathode and a PEO polymer as an electrolyte, the energy density of 160 Wh kg^{-1} is obtained with 40%–75% using the capacity [121]. However, the operation is limited to an elevated temperature of 50–93°C. At 20°C, the energy density of 82 Wh kg^{-1} was obtained using a liquid electrolyte containing dimethyl sulfoxide (DMSO) [122]. The energy density increases to 303 Wh kg^{-1}, when 2,5-dimercapto-1,3,4-thiadiazole (DMcT) is used as a positive electrode at 25°C [119]. Due to its high stability of cyclic work over a wide temperature range, DMcT has been extensively studied, but its theoretical capacity is only 362 mAh g^{-1}. Similar to the cathode of elemental sulfur, the electrically insulating character of DMcT and the high solubility of its discharging product in the liquid electrolyte result in poor cyclic operation. The electrical conductivity is improved by using a conductive polymeric organic sulfide, poly (dithiodianilin) as the cathode material. Coupled skeletons are preserved during cyclic operation and provide high electrical conductivity. The mass fraction of conductive skeletons reduces the capacity to 270 mAh g^{-1}. Interpolymer disulfide linkages between individual polymer chains cannot offer high incorporation efficiency, and consequently conductive polymeric organic sulfides show unsatisfactory cyclic function. Despite major research efforts in the field of organic sulfides, their use as positive electrodes was not particularly successful [119, 123–132]. Their impracticality makes them less attractive than elemental sulfur.

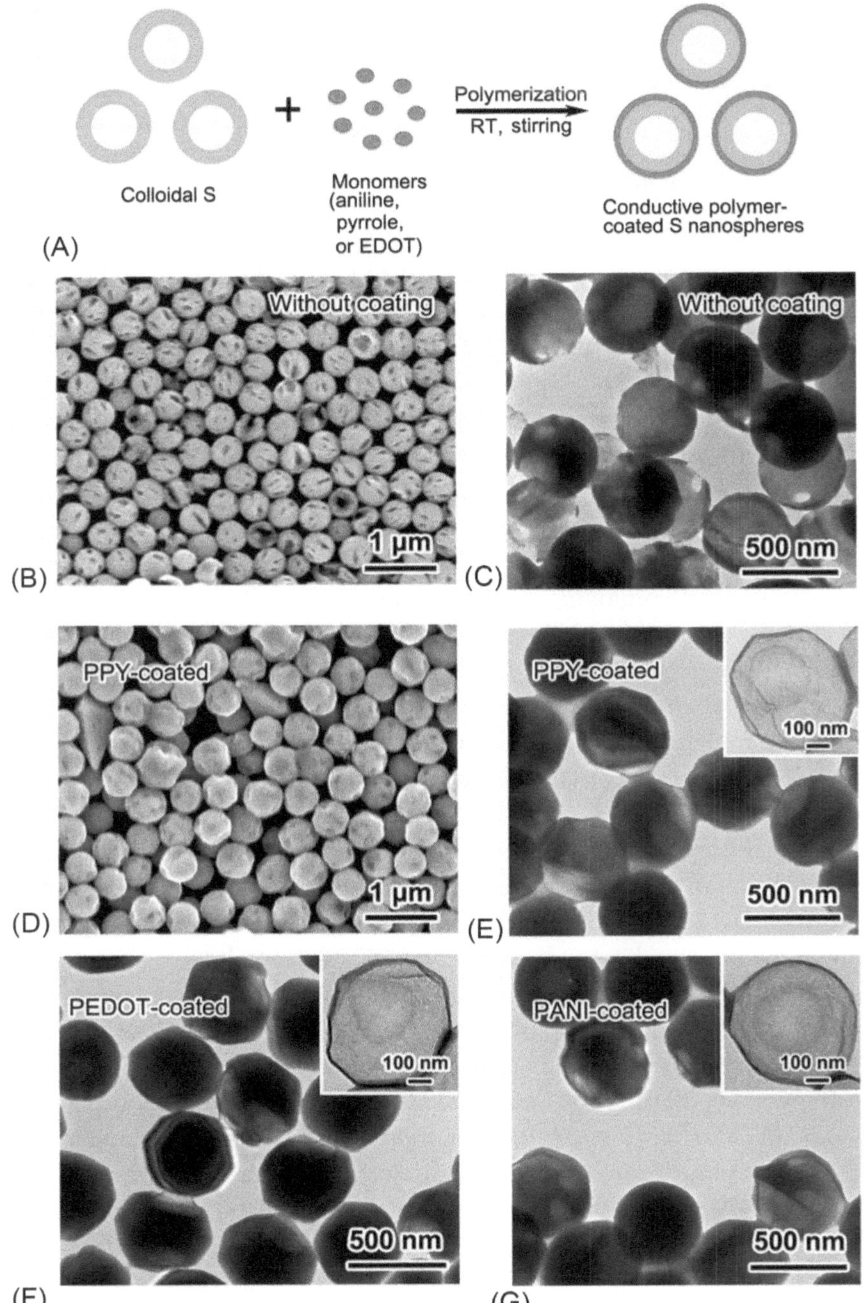

Fig. 19 (A) Schematic illustration of the fabrication process of conductive polymer-coated hollow sulfur nanospheres. RT, room temperature. (B, D) Scanning electron microscopy (SEM) and (C, E) transmission electron microscopy (TEM) images of the hollow sulfur nanospheres (B, C) before and (D, E) after coating with polypyrrole (PPY). (F, G) TEM images of the (F) poly(3,4-ethylenedioxythiophene) (PEDOT)-and (G) polyaniline (PANI)-coated hollow sulfur nanospheres. Insets in E–G: TEM images of the PPY, PEDOT, and PANI shell after dissolving sulfur with toluene. *Taken from W. Li, Q. Zhang, G. Zheng, Z.W. Seh, H. Yao, Y. Cui, Understanding the role of different conductive polymers in improving the nanostructured sulfur cathode performance, Nano Lett. 13(11) (2013) 5534–5540.*

5.4 HIGH-LOADING SULFUR CATHODES

A high-loading electrode is essential for establishing high energy density lithium-sulfur (Li-S) batteries, but it is confronted with critical challenges.

A comparative analysis of the structural cathodes with increasing sulfur loading, presented in [133], provides insights into the development of advanced sulfur cathodes with high electrochemical performance and attractive active-material loading. It was found that the insulating sulfur core may form within the active-material fillings and thereby reduce the initial active-material utilization. The possible solution was to channel the dissolved polysulfides to activate the insulating sulfur clusters. There are many publications on Li-S batteries employing regular-loading sulfur cathodes (sulfur loading $<2\,mg\,cm^{-2}$), but new challenges such as increasing polarization and cell resistance that arise with high-loading sulfur cathodes should be tackled.

A core-shell structural sulfur cathode was elaborated to study the feasibility of holding a high-loading sulfur core within a carbon-shell electrode configuration. This concept used the unique material chemistry of sulfur rather than restricting it. The formation and diffusion of polysulfides were in charge of activating the high-loading sulfur core and improving the electrochemical utilization of sulfur. The carbon-shell electrode provided the high-loading sulfur core with fast ion and electron transport and stabilized the active material within the structural cathode configuration. As a result, the sulfur-carbon core-shell cathode effectively utilized the stabilized sulfur core within the carbon-shell electrode, demonstrating an overall boost in the electrochemical utilization and polysulfide retention. The core-shell cathodes with high sulfur loadings of $4.0–30.0\,mg\,cm^{-2}$ exhibited high cycle stability at various cycling rates (0.05C to 0.5C rates). For example, the core-shell cathode with a $4\,mg\,cm^{-2}$ sulfur loading showed high electrochemical utilization of sulfur of above 96% with stable electrochemical cyclability for over 100 cycles at a 0.2C rate. The high-loading core-shell cathodes with 20 and $30\,mg\,cm^{-2}$ sulfur loadings attained peak discharge capacities of, respectively, 870 and $780\,mAh\,g^{-1}$ at a 0.2C rate. Such high electrochemical utilization facilitated a high areal capacity of $17–23\,mAh\,cm^{-2}$.

Another strategy was presented in [134], proposing blending sulfur with a transition metal sulfide (TiS_2 and MoS_2) to form dense composite cathodes with enhanced conductivity. An improvement was obtained in both the initial capacity from sulfur utilization ($800\,mAh\,g^{-1}$ based on sulfur content), the coulombic efficiency (>96%) and also in cycle life upon blending with the metal sulfide. High sulfur loadings (>12 mg cm^{-2} or 6 mAh cm^{-2} per side) were shown to display high sulfur utilization in Li-S cells containing the metal sulfide blends, either with or without coatings over the sulfur cathode.

According to [135], to obtain an energy density advantage, the performance degradation of high sulfur-loading cathodes became an urgent problem to be solved. Additionally, the volumetric capacities of high sulfur-—loading cathodes are still at a low level compared with their areal capacities. Aiming at these issues, a two-dimensional carbon yolk-shell nanosheet was developed (Fig. 20) to construct a novel self-supporting sulfur cathode. The cathode with high-sulfur loading of 5 mg cm^{-2} and sulfur content of 73 wt% delivered an excellent rate performance and cycling stability. It also provided a favorable balance between the areal (5.7 mAh cm^{-2}) and volumetric (1330 mAh cm^{-3}) capacities. It was found that an areal capacity of 11.4 mAh cm^{-2} can be further achieved by increasing the sulfur loading from 5 to 10 mg cm^{-2}.

Studies described in [136] used a conventional pure sulfur cathode. It was fabricated by tape casting the mixed active-material paste onto an Al foil current collector and then drying the *N*-methyl-2-pyrolidone in an air oven for 48 h at 50°C. The pure sulfur cathodes consisted of 75 wt% sulfur, 15 wt%

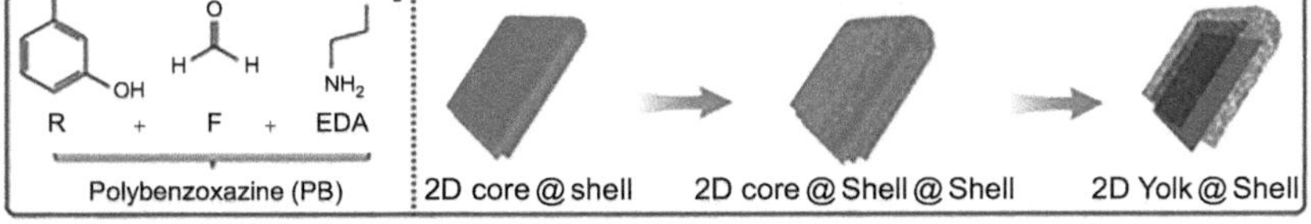

Fig. 20 Schematic illustration for the synthesis of G@HMCN. Two-step coating of GO with SiO_2 and porous carbon precursor PB/SiO_2, followed by the carbonization and etching of SiO_2 to produce G@HMCN. Tetraethylorthosilicate (TEOS) was employed as the SiO_2 precursor, and resorcinol (R), formaldehyde (F), and ethylenediamine (EDA) were employed as the PB precursors. *Taken from F. Pei, L. Lin, D. Ou, Z. Zheng, S. Mo, X. Fang, N. Zheng, Self-supporting sulfur cathodes enabled by two-dimensional carbon yolk-shell nanosheets for high-energy-density lithium-sulfur batteries, Nat. Commun. 8 (2017) 482. doi:10.1038/s41467-017-00575-8.*

Super P (TIMCAL), and 10 wt% polyvinylidene fluoride (PVdF, Kureha). The high sulfur content cathodes contained 80 wt% sulfur with 10 wt% Super P and 10 wt% PVdF. The pure sulfur cathodes used pure sulfur powders instead of the sulfur-based nanocomposites as the active material. The different sulfur-loading cathodes were prepared by controlling the thickness of the active-material coating (1.5 mg cm^{-2}: 18.1 μm; 3.0 mg cm^{-2}: 23.0 μm; 6.3 mg cm^{-2}: 49.1 μm). The highest sulfur loading and content of a pure sulfur cathode used in this work was 6.3 mg cm^{-2} and 80 wt% (S/C ratio ~ 7). As a reference, a high sulfur content of 78 wt% is still achieved after including the weight of the SWCNT layer. The pure sulfur cathode mated with a custom single-wall carbon nanotube (SWCNT)-modulated separator. The cell employing pure sulfur cathode mating with the SWCNT-modulated separator exhibited a high discharge capacity of 1132 mA h g^{-1} with a low-capacity fade rate of 0.18% per cycle after 300 cycles.

In [137] a free-standing poached egg–shaped architecture was presented through a facile template-supported vacuum-filtration strategy and it was employed as an efficient sulfur host for Li-S batteries. This unique architecture guaranteed an effective encapsulation of the "sulfur yolk" inside the fully vacuum-sealed framework, effectively limiting the active material loss and polysulfide diffusion. The conductive and porous framework served as an interlinked electron pathway and electrolyte channel, greatly facilitating fast electric/ionic transport along with active material reactivation and reutilization during cycling. A high peak discharge capacity (1200 mA h g^{-1}), a low capacity-fade rate (0.09% cycle^{-1}) for 500 cycles, and excellent rate capability (C/5–1C rates) were accomplished. With such an advantageous architecture, the sulfur loading was increased to 32 mg cm^{-2} to achieve an areal capacity of up to 16 mA h cm^{-2} (Fig. 21).

According to [138] the large volume change during cycles, extreme low electronic/ionic conductivity of sulfur and lithium sulfide, and high solubility of high-order polysulfide (PS) in electrolyte are major disadvantages of classic Li-S cells. The current strategy to mitigate these three issues is to physically and/or chemically encapsulate the sulfur into a porous carbon host (PCH), which is only suitable for a low S loading cathode. The large volume change during the charge-discharge process especially in a thicker electrode with high areal S loading will detach the carbon/Li_2S particles apart from each other, resulting in fast capacity decay. Although incorporation of a large amount of S into 3D carbon paper (CP) can maintain the electrode integration, it cannot effectively immobilize soluble PS. During studies, the PS was chemically bonded to both doped PCH and 3D doped CP current collector

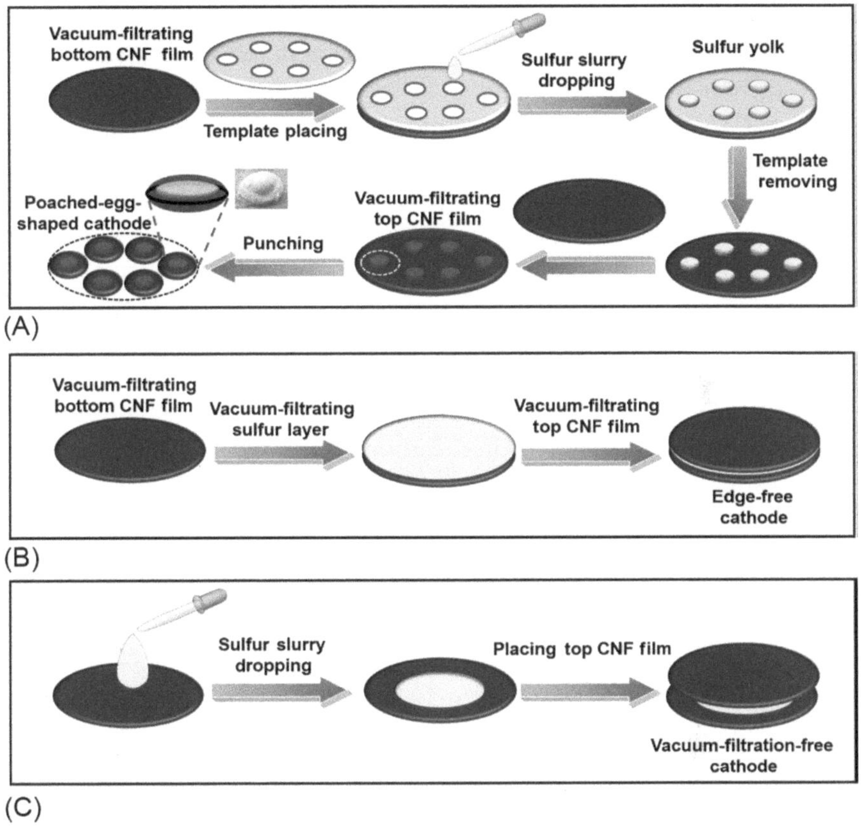

Fig. 21 Fabrication processes of the (A) poached-egg shaped, (B) edge-free, and (C) vacuum filtration–free cathodes. *Taken from L. Luo, A. Manthiram, Rational design of high-loading sulfur cathodes with a poached-egg-shaped architecture for long-cycle lithium–sulfur batteries, ACS Energy Lett. 2(10) (2017) 2205–2211.*

to suppress the shuttle reaction of PS, and also to maintain the electronic/ionic connection during prolonged cycles. The S, N doped CP/doped PCH/S cathode with high areal sulfur loading of 9.0 mg cm^{-2} provided a high capacity of 1013 mA h g^{-1} with a slow capacity fading rate of 0.074% per cycle for 300 cycles, presenting one of the best cycling stabilities at similar S loading reported to date. Chemical bonding of PS to CP current collector can more effectively enhance the cycling stability of a high sulfur-loading cathode, compared to the current method by chemical bonding of PS to PCH. This sulfur cathode, by filling a large amount of S/C composites into a functionalized CP matrix, can allow realization of high-energy lithium-sulfur batteries.

In [139] a Li-S battery with a high areal capacity is proposed by a systematic strategy incorporating two approaches as follows: (1) a hierarchically porous carbon host containing graphene (G), mesoporous carbon (MPC), and super P (SP) diminishes polysulfide migration and guarantees fast electron and ion transport in a thick cathode; (2) a glass-fiber (GF) membrane severs as the electrolyte reservoir to prevent the short circuit resulted from the deficiency of liquid electrolyte. With these methods, the Li-S batteries with an ultrahigh sulfur loading of 13 mg cm^{-2} provided a high areal capacity of 14.3 mA h cm^{-2} (1099 mA h g^{-1}) at the first cycle and stable cycling performance with a reversible capacity of 628 mA h g^{-1} (8.16 mA h cm^{-2}) after 75 cycles at 0.1 C.

Reference [140] presented a composite paper electrode consisting of hollow Co_3S_4 polyhedra, activated carbon nanofibers (ACNF), and pure sulfur powder for Li-S batteries. The hollow Co_3S_4 polyhedra with a porous shell structure provided large spaces to accommodate the polysulfides, possessed ideal polar chemisorptive capability for immobilizing the polysulfide species, and had a higher conductivity than many polar metal oxides, facilitating fast reaction kinetics with the Li-S cells. The composite electrode exhibited high sulfur utilization and maintained high cycling stability in Li-S cells with a high sulfur areal mass. A high capacity of 953 mAh g^{-1} at a 1C rate could be achieved with a sulfur loading of 2.5 mg cm^{-2}, and the capacity remained at 610 mAh g^{-1} after 450 cycles. Even with a high sulfur loading of 13.5 mg cm^{-2}, the composite electrode gave a high areal capacity of 13 mAh cm^{-2} at a 0.3C rate.

5.5 OTHER POSITIVE ELECTRODE MATERIALS

Metal sulfides (M-S) were first used, in order to replace elemental sulfur, as positive electrodes for high temperature (over 400°C) lithium-sulfur batteries [141]. Many M-S's (M = Fe, Co, Ni, etc.) are tested as positive electrodes, because to some extent they solve the problems of corrosion and variability associated with the use of elemental sulfur in high-temperature batteries. The operating temperature is limited to 95–105°C using a polymer composite electrolyte to replace molten electrolyte salts and a magnesium oxide separator [142, 143]. Distributable through commercialization of lithium-ion batteries, sulfides of transition metals and lithium sulfide/binary (divalent) metals Li-MS (M = Fe [144, 145], Co [146], Cu ([147, 148], Ti [149] and Mo [150]) were intensively tested as positive electrodes for lithium-sulfur batteries operating at room temperature.

M serves as a conductive material to replace conventional carbonaceous materials. The Cu_2S nanowire structures were directly grown on the Cu substrate and were used as a positive electrode that provides a precharge capacity of $400\,mAh\,g^{-1}$ and maintains more than 50% of this reversible capacity after 100 cycles [151]. By comparison, Ni_3S_2 provides a higher initial capacity of $430\,mAh\,g^{-1}$ and retains more than 80% of the reversible capacity [152]. A high initial capacity of $1000\,mAh\,g^{-1}$ is obtained for Co_9S_8 nanotubes, but only 37% of this capacity is maintained [153], while SnS_2 nanowires can provide a very stable reversible capacity above $500\,mAh\,g^{-1}$ after 50 cycles [154]. Other positive materials, such as V_2O_5-S transition metal oxide [125] and inorganic phosphazene disulfide polymer $[(NPS_2)_3]_n$ [155], are reported to reduce cell resistance and improve high-current efficiency for lithium-sulfur batteries.

Paper [156] presents a method for obtaining high-quality Li_2S crystalline nanomaterials of an average particle size of about 55 nm and coated with Li_3PS_4 to form a nanoscale lithium-layered composite Li_2S @ Li_3PS_4. Next, this material was used to produce a nano-Li_2S @ Li_3PS_4/graphene aerogel in a simple liquid-infiltration/evaporation process, used directly as a composite cathode without a metal substrate for a lithium-sulfur battery. This composite provided a high discharging capacity of $934.4\,mAh\,g^{-1}$ in the initial cycle and maintains a capacity of $485.5\,mAh\,g^{-1}$ after 100 cycles at a current load of 0.1 C. Furthermore, the composite exhibited a significantly lower potential barrier (~2.40 V) and overpotential compared to previous reports, indicating that Li_2S only needs a small amount of energy to be activated. Excellent electrochemical properties resulted from the small size of Li_2S particles and the presence of a superconducting Li_3PS_4 coating layer, which could shorten Li and electron diffusion paths, improve ionic conductivity, and delay the dissolution of polysulfides to some extent in the electrolyte.

Paper [157] presented an interesting strategy consisting of the formation of an adsorption-blocking polysulfide (PAL) in situ on the surface of the cathode, formed in situ to control the transfer of polysulfides and promote the stability of cyclic work in Li-S batteries. PAL consists of La_2S_3, which are able to chemically adsorb polysulfides through strong interaction of La-S bonds and S-S bonds and build an effective barrier against the escape of sulfur. In addition, La_2S_3 are able to inhibit the crystallization of Li_2S and facilitate the transfer of ions, which contributes to reducing the internal electrical resistance of the battery. In addition, the by-product $LiNO_3$ simultaneously forms a stable solid intermediate phase between the anode and the electrolyte to further inhibit polysulfide transfer. Thanks to this method, Li-S batteries achieved high cyclic operation stability, with a reduction in capacity of only 0.055% from the 10th cycle.

Reference [158] describes a composite cathode sulfur/polyacrylonitrile/reduced graphene oxide, synthesized in the self-assembly process. The electrical conductivity of the composite increased from $\sim 10^{-12}$ to $\sim 10^{-4}$ - $S\,cm^{-1}$. This composite provides a high capacity of 700 mAh $(gS)^{-1}$ even at a load of 2 C ($3.4\,A\,g^{-1}$). CS bonding in the composite was retained even after intensive cyclic loading. In paper [159] a new cathode was presented for a Li-S cell using powder sulfur deposited in a double coating of magnesium dioxide and graphene oxide (S @ MnO_2 @ GO) with improved current loading capacity and for cyclic work. To create this structure, a reaction using a sulfur-reducing $KMnO_4$ was used to generate in situ MnO_2, which covers the surface of excess sulfur. The resulting MnO_2 with a honeycomb-like morphology provided excellent storage space for polysulfides. The outer GO layer was created to block the open pores of MnO_2, thus minimizing the tendency of polysulfides to dissolve in the electrolytes. GO significantly improved the electrical conductivity of the sulfur cathode, and the structure of S @ MnO_2 @ GO showed excellent ability to load and long life. In the study in [160], a new type of elastic porous layer of carbon nanoparticles modified with graphene nanoparticles and ultrafine polar TiO_2 nanoparticles was presented as a matrix to keep sulfur in the Li-S battery cathode. The excellent structure of the layer enabled high-performance sulfur dispersion in order to obtain a large capacity and a high ability to separate lithium polysulfides, resulting in a long service life. This cathode showed a high initial discharge capacity of 1501 mAh g^{-1} at 0.1 C load and the ability to carry high current loads, resulting in a 668 mAh g^{-1} specific charge at 5 C load, as well as extended cyclic stability. Reference [161] showed the covalent grafting path used to produce a composite of homogeneous hollow carbon nanocycle (HCN) and trithiocyanuric acid (TTCA) vulcanized in HCN pores. This composite demonstrated the high use of vulcanized TTCA by HCN with a large area of $2330\,m^2\,g^{-1}$ and a high pore volume and covalent bonds with sulfur, effectively inhibiting the dissolution of polysulfides. The capacity of the first discharge of the composite reached 1430 mAh g^{-1} at a load of 0.1 C and $1227\,mAh\,g^{-1}$ at a load of 0.2 C.

As considered in [162, 163], Li/polysulfide batteries can be used to increase the areal sulfur loading and enhance both energy density and power density.

According to [17], the development of Li/polysulfide cells mainly concerned the electrode design. Tin-doped indium oxide–decorated CNF paper, Pt/graphene, nitrogen-doped graphene paper, and TiO_2 nanowire/graphene hybrid membrane were used as bifunctional polysulfide immobilizers and current collectors into this system.

According to [164], a second protection, such as a graphene coating on the electrolyte/separator interface, can further block the diffusion of polysulfides.

As shown in [163, 165], such a configuration can further derive Li/polysulfide redox flow batteries with further improved energy densities and power densities. However, according to [3], with the employment of polysulfides and aggravation of flow convection, the shuttle effect becomes much more severe.

As verified in [163, 165, 166], the addition of a large ratio of lithium salt (LiTf and LiTFSI) or percolating nanoscale conductor suspension (Ketjen Black) in the electrolyte enabled the cycling of polysulfide catholyte in a cathode-flow mode to improve the capacity utilization.

According to [7], relatively little emphasis was placed on dealing with the volumetric expansion of sulfur during lithiation, leading to cracking and fracture of the protective shell. A sulfur-TiO_2 yolk-shell nanoarchitecture was designed with internal void space to accommodate the volume expansion of sulfur, resulting in an intact TiO_2 shell to minimize polysulfide dissolution. An initial specific capacity of 1030 mAh g^{-1} at 0.5 C and coulombic efficiency of 98.4% over 1000 cycles were achieved. Most importantly, the capacity decay after 1000 cycles was as small as 0.033% per cycle.

Reference [167] reported a three-dimensional (3D) electrode structure to achieve both sulfur physical encapsulation and polysulfides binding simultaneously. The electrode was based on hydrogen reduced TiO_2 with an inverse opal structure that is highly conductive and robust toward electrochemical cycling. The relatively enclosed 3D structure provided an ideal architecture for sulfur and polysulfide confinement. The openings at the top surface allowed sulfur infusion into the inverse opal structure. Chemical tuning of the TiO_2 composition through hydrogen reduction enhanced the specific capacity and cyclability of the cathode. With such TiO_2 encapsulated sulfur structure, the sulfur cathode delivered a high specific capacity of 1100 mAh g^{-1} in the beginning, with a reversible capacity of 890 mAh g^{-1} after 200 cycles of charge/discharge at a C/5 rate. The coulombic efficiency was around 99.5% during cycling.

According to [168], the conductive Magnéli phase Ti_4O_7 is a highly effective matrix to bind with sulfur species. Compared with the TiO_2-S, the Ti_4O_7-S cathodes exhibited higher reversible capacity and improved cycling performance. They delivered high specific capacities at various C-rates (1342, 1044, and 623 mAh g^{-1} at 0.02, 0.1, and 0.5 C, respectively) and remarkable capacity retention of 99% (100 cycles at 0.1 C). The superior properties of Ti_4O_7-S were attributed to the strong adsorption of sulfur species on the low-coordinated Ti sites of Ti_4O_7.

Reference [11] reported the encapsulation of Li_2S cathodes using two-dimensional layered transition metal disulfides that possessed a combination of high conductivity and strong binding with Li_2S/Li_2S_n species. Using titanium disulfide as an encapsulation material, a high specific capacity of 503 mAh g^{-1} Li_2S under high C-rate conditions (4C) was obtained, as well as high areal capacity of 3.0 mAh cm^{-2} under high mass-loading conditions (5.3 mgLi_2S cm^{-2}).

During studies described in [169], a two-dimensional layered molybdenum disulfide (MoS_2) was used to show the electrochemical selectivity of edge vs. terrace sites for Li-S batteries and hydrogen evolution reaction (HER). Lithium sulfide (Li_2S) nanoparticles decorated along the edges of the MoS_2 nanosheet vs. terrace, confirming the strong binding energies between Li_2S and the edge sites and guiding the improved electrode design for Li-S batteries.

REFERENCES

[1] Z. Liu, W. Fu, C. Liang, Chapter 24 Lithium–sulfur batteries, in: C. Daniel, J.O. Besenhard (Eds.), Handbook of Battery Materials, second ed., Wiley-VCH Verlag GmbH & Co. KGaA, 2011.

[2] X.L. Ji, K.T. Lee, L.F. Nazar, A highly ordered nanostructured carbon-sulphur cathode for lithium-sulphur batteries, Nat. Mater. 8 (2009) 500–506.

[3] J.C. Guo, Y.H. Xu, C.S. Wang, Sulfur-impregnated disordered carbon nanotubes cathode for lithium-sulfur batteries, Nano Lett. 11 (2011) 4288–4294.

[4] S. Xin, L. Gu, N.-H. Zhao, Y.-X. Yin, L.-J. Zhou, Y.-G. Guo, L.-J. Wan, Smaller sulfur molecules promise better lithium-sulfur batteries, J. Am. Chem. Soc. 134 (2012) 18510–18513.

[5] L. Wang, D. Wang, F.X. Zhang, J. Jin, Interface chemistry guided long-cycle-life Li-S battery, Nano Lett. 13 (2013) 4206–4211.

[6] Q. Pang, D. Kundu, M. Cuisinier, L.F. Nazar, Surface-enhanced redox chemistry of polysulphides on a metallic and polar host for lithium-sulphur batteries, Nat. Commun. 5 (4759) (2014).

[7] Z. Wei Seh, W. Li, J.J. Cha, G. Zheng, Y. Yang, M.T. McDowell, P.-C. Hsu, Y. Cui, Sulphur-TiO_2 yolk-shell nanoarchitecture with internal void space for long-cycle lithium-sulphur batteries, Nat. Commun. 4 (2013) 1331.

[8] W.D. Zhou, Y.C. Yu, H. Chen, F.J. DiSalvo, H.D. Abruña, Yolk-shell structure of polyaniline-coated sulfur for lithium-sulfur batteries, J. Am. Chem. Soc. 135 (2013) 16736–16743.

[9] Y. Yang, G. Yu, J.J. Cha, H. Wu, M. Vosgueritchian, Y. Yao, Z. Bao, Y. Cui, Improving the performance of lithium-sulfur batteries by conductive polymer coating, ACS Nano 5 (2011) 9187–9193.
[10] H.W. Chen, C. Wang, Y. Dai, S. Qiu, J. Yang, W. Lu, L. Chen, Rational design of cathode structure for high rate performance lithium-sulfur batteries, Nano Lett. 15 (2015) 5443–5448.
[11] Z.W. Seh, J.H. Yu, W. Li, P.-C. Hsu, H. Wang, Y. Sun, H. Yao, Q. Zhang, Y. Cui, Two-dimensional layered transition metal disulphides for effective encapsulation of high-capacity lithium sulphide cathodes, Nat. Commun. 5 (2014) 5017.
[12] Y.S. Su, A. Manthiram, Lithium-sulphur batteries with a microporous carbon paper as a bifunctional interlayer, Nat. Commun. 3 (2012) 1166.
[13] G.M. Zhou, S. Pei, L. Li, D.-W. Wang, S. Wang, K. Huang, L.-C. Yin, F. Li, H.-M. Cheng, A graphene-pure-sulfur sandwich structure for ultrafast, long-life lithium-sulfur batteries, Adv. Mater. 26 (2014) 625–631.
[14] Y.J. Li, J.M. Fan, M.S. Zheng, Q.F. Dong, A novel synergistic composite with multi-functional effects for high-performance Li-S batteries, Energy Environ. Sci. 9 (2016) 1998–2004.
[15] S. Rehman, X. Gu, K. Khan, N. Mahmood, W. Yang, X. Huang, S. Guo, Y. Hou, 3D vertically aligned and interconnected porous carbon nanosheets as sulfur immobilizers for high performance lithium-sulfur batteries, Adv. Energy Mater. 6 (2016) 1502518.
[16] F. Pei, T. An, J. Zang, X. Zhao, X. Fang, M. Zheng, Q. Dong, N. Zheng, From hollow carbon spheres to N-doped hollow porous carbon bowls: rational design of hollow carbon host for Li-S batteries, Adv. Energy Mater. 6 (2016).
[17] M. Liu, X. Qin, Y.-B. He, B. Li, F. Kang, Recent innovative configurations in high-energy lithium–sulfur batteries, J. Mater. Chem. A 5 (2017) 5222–5234.
[18] M.L.B. Rao, Organic electrolyte cells, US patent 3,413,154, 1968.
[19] D.A. Nole, V. Moss, Battery employing lithium-sulphur electrodes with non-aqueous electrolyte, US Patent 3,532,543, 1970.
[20] B.H. Jeon, J.H. Yeon, K.M. Kim, I.J. Chung, Preparation and electrochemical properties of lithium–sulfur polymer batteries, J. Power Sources 109 (2002) 89–97.
[21] S.E. Cheon, K.S. Ko, J.H. Cho, S.W. Kim, E.Y. Chin, H.T. Kim, Rechargeable lithium sulfur battery I. Structural change of sulfur cathode during discharge and charge, J. Electrochem. Soc. 150 (2003) A796–A799.
[22] S.E. Cheon, K.S. Ko, J.H. Cho, S.W. Kim, E.Y. Chin, H.T. Kim, Rechargeable lithium sulfur battery II. Rate capability and cycle characteristics, J. Electrochem. Soc. 150 (2003) A800–A805.
[23] J. Hassoun, B. Scrosati, A high-performance polymer tin sulfur lithium ion battery, Angew. Chem. Int. Ed. 49 (2010) 2371–2374.
[24] X. Zhao, D.S. Kim, H.J. Ahn, K.W. Kim, C.S. Jin, J.H. Ahn, Effect of preparation parameters of sulfur cathodes on electrochemical properties of lithium sulfur battery, J. Korean Electrochem. Soc. 13 (2010) 169–174.
[25] H.S. Ryu, H.J. Ahn, K.W. Kim, J.H. Ahn, K.K. Cho, T.H. Nam, J.U. Kim, G.B. Cho, Discharge behavior of lithium/sulfur cell with TEGDME based electrolyte at low temperature, J. Power Sources 163 (2006) 201–206.
[26] W.K. Wang, Y. Wang, Y.Q. Huang, C.J. Huang, Z.B. Yu, H. Zhang, A.B. Wang, K.G. Yuan, The electrochemical performance of lithium–sulfur batteries with $LiClO_4$ DOL/DME electrolyte, J. Appl. Electrochem. 40 (2010) 321–325.
[27] L.X. Yuan, J.K. Feng, X.P. Ai, Y.L. Cao, S.L. Chen, H.X. Yang, Improved dischargeability and reversibility of sulfur cathode in a novel ionic liquid electrolyte, Electrochem. Commun. 8 (2006) 610–614.
[28] A. Swiatkowski, Some uses of carbon materials in electrochemistry, Przem. Chem. (2017), https://doi.org/10.15199/62.2017.1.7.

[29] D.-W. Wang, Q. Zeng, G. Zhou, L. Yin, F. Li, H.-M. Cheng, I.R. Gentle, G.Q.M. Lu, Carbon–sulfur composites for Li–S batteries: status and prospects, J. Mater. Chem. A 1 (2013) 9382.
[30] S.E. Cheon, S.S. Choi, J.S. Han, Y.S. Choi, B.H. Jung, H.S. Lim, Capacity fading mechanisms on cycling a high-capacity secondary sulfur cathode, J. Electrochem. Soc. 151 (2004) A2067–A2073.
[31] J.L. Wang, L. Liu, Z.J. Ling, J. Yang, C.R. Wan, C.Y. Jiang, Polymer lithium cells with sulfur composites as cathode materials, Electrochim. Acta 48 (2003) 1861–1867.
[32] B. Zhang, C. Lai, Z. Zhou, X.P. Gao, Preparation and electrochemical properties of sulfur–acetylene black composites as cathode materials, Electrochim. Acta 54 (2009) 3708–3713.
[33] V.S. Kolosnitsyn, E.V. Karaseva, N.A. Amineva, G.A. Batyrshina, Cycling lithium–sulfur batteries, Russ. J. Electrochem. 38 (2002) 329–331.
[34] F. Wu, S.X. Wu, R.J. Chen, S. Chen, G.Q. Wang, Electrochemical performance of sulfur composite cathode materials for rechargeable lithium batteries, Chin. Chem. Lett. 20 (2009) 1255–1258.
[35] R. Elazari, G. Salitra, Y. Talyosef, J. Grinblat, C. Scordilis-Kelley, A. Xiao, J. Affinito, D. Aurbach, Morphological and structural studies of composite sulfur electrodes upon cycling by HRTEM, AFM and Raman spectroscopy, J. Electrochem. Soc. 157 (2010) A1131–A1138.
[36] Y.S. Choi, S. Kim, S.S. Choi, J.S. Han, J.D. Kim, S.E. Jeon, B.H. Jung, Effect of cathode component on the energy density of lithium–sulfur battery, Electrochim. Acta 50 (2004) 833–835.
[37] S.C. Han, M.S. Song, H. Lee, H.S. Kim, H.J. Ahn, J.Y. Lee, Effect of multiwalled carbon nanotubes on electrochemical properties of lithium/sulfur rechargeable batteries, J. Electrochem. Soc. 150 (2003) A889–A893.
[38] W. Zheng, Y.W. Liu, X.G. Hu, C.F. Zhang, Novel nanosized adsorbing sulfur composite cathode materials for the advanced secondary lithium batteries, Electrochim. Acta 51 (2006) 1330–1335.
[39] J.J. Chen, X. Jia, Q.J. She, C. Wang, Q.A. Zhang, M.S. Zheng, Q.F. Dong, The preparation of nano-sulfur/MWCNTs and its electrochemical performance, Electrochim. Acta 55 (2010) 8062–8066.
[40] Y.J. Choi, K.W. Kim, H.J. Ahn, J.H. Ahn, Improvement of cycle property of sulfur electrode for lithium/sulfur battery, J. Alloys Compd. 449 (2008) 313–316.
[41] L.X. Yuan, H.P. Yuan, X.P. Qiu, L.Q. Chen, W.T. Zhu, Improvement of cycle property of sulfur-coated multi-walled carbon nanotubes composite cathode for lithium/sulfur batteries, J. Power Sources 189 (2009) 1141–1146.
[42] E. Dujardine, T.W. Ebbesen, H. Hiura, K. Tanigaki, Capillarity and wetting of carbon nanotubes, Science 265 (1994) 1850–1852.
[43] C. Wang, J.J. Chen, Y.N. Shi, M.S. Zheng, Q.F. Dong, Preparation and performance of a core–shell carbon/sulfur material for lithium/sulfur battery, Electrochim. Acta 55 (2010) 7010–7015.
[44] S. Dörfler, M. Hagen, H. Althues, J. Tübke, S. Kaskel, M.J. Hoffmann, High capacity vertical aligned carbon nanotube/sulfur composite cathodes for lithium–sulfur batteries, Chem. Commun. 48 (2012) 4097–4099.
[45] Y.J. Choi, Y.D. Chung, C.Y. Baek, K.W. Kim, H.J. Ahn, J.H. Ahn, Effects of carbon coating on the electrochemical properties of sulfur cathode for lithium/sulfur cell, J. Power Sources 184 (2008) 548–552.
[46] J. Wang, S.Y. Chew, Z.W. Zhao, S. Ashraf, D. Wexler, J. Chen, S.H. Ng, S.L. Chou, H.K. Liu, Sulfur–mesoporous carbon composites in conjunction with a novel ionic liquid electrolyte for lithium rechargeable batteries, Carbon 46 (2008) 229–235.

[47] C. Lai, X.P. Gao, B. Zhang, T.Y. Yan, Z. Zhou, Synthesis and electrochemical performance of sulfur/highly porous carbon composites, J. Phys. Chem. C 113 (2009) 4712–4716.
[48] C.D. Liang, N.J. Dudney, J.Y. Howe, Hierarchically structured sulfur/carbon nanocomposite material for high-energy lithium battery, Chem. Mater. 21 (2009) 4724–4730.
[49] B. Zhang, X. Qin, G.R. Li, X.P. Gao, Enhancement of long stability of sulfur cathode by encapsulating sulfur into micropores of carbon spheres, Energy Environ. Sci. 3 (2010) 1531–1537.
[50] M. Sevilla, R. Foulston, R. Mokaya, Superactivated carbide-derived carbons with high hydrogen storage capacity, Energy Environ. Sci. 3 (2010) 223–227.
[51] Y. Zhang, L.Z. Wang, A.Q. Zhang, Y.H. Song, X.F. Li, H. Feng, X.B. Wu, P.P. Du, Novel V_2O_5/S composite cathode material for the advanced secondary lithium batteries, Solid State Ionics 181 (2010) 835–838.
[52] M.S. Song, S.C. Han, H.S. Kim, J.H. Kim, K.T. Kim, Y.M. Kang, H.J. Ahn, S.X. Dou, J.Y. Lee, Effects of nanosized adsorbing material on electrochemical properties of sulfur cathodes for Li/S secondary batteries, J. Electrochem. Soc. 151 (2004) A791–A795.
[53] N. Jayaprakash, J. Shen, S.S. Moganty, A. Corona, L.A. Archer, Porous hollow carbon·sulfur composites for high-power lithium-sulfur batteries, Angew. Chem. Int. Ed. Eng. 50 (26) (2011) 5904–5908.
[54] J. Kim, D.-J. Lee, H.-G. Jung, Y.-K. Sun, J. Hassoun, B. Scrosati, An advanced lithium-sulfur battery, Adv. Funct. Mater. 23 (8) (2013) 1076–1080.
[55] W. Li, G. Zheng, Y. Yang, Z.W. Seh, N. Liu, Y. Cui, High-performance hollow sulfur nanostructured battery cathode through a scalable, room temperature, one-step, bottom-up approach, PNAS 110 (18) (2013) 7148–7153.
[56] Y. Zhang, X.B. Wu, H. Feng, L.Z. Wang, A.Q. Zhang, T.C. Xia, H.C. Dong, Effect of nanosized Mg0.8Cu0.2O on electrochemical properties of Li/S rechargeable batteries, Int. J. Hydrog. Energy 34 (2009) 1556–1559.
[57] Y.J. Choi, B.S. Jung, D.J. Lee, J.H. Jeong, K.W. Kim, H.J. Ahn, K.K. Cho, H.B. Gu, Electrochemical properties of sulfur electrode containing nano Al2O3 for lithium/sulfur cell, Phys. Scr. 129 (2007) 62–65.
[58] X.M. He, J.G. Ren, L. Wang, W.H. Pu, C.R. Wan, C.Y. Jiang, Electrochemical characteristics of sulfur composite cathode for reversible lithium storage, Ionics 15 (2009) 477–481.
[59] Y. Yang, M.T. McDowell, A. Jackson, J.J. Cha, S.S. Hong, Y. Cui, New nanostructured Li2S/silicon rechargeable battery with high specific energy, Nano Lett. 10 (2010) 1486–1491.
[60] J. Wang, Y.W. Wang, X.M. He, J.G. Ren, C.Y. Jiang, C.R. Wan, Electrochemical characteristics of sulfur composite cathode materials in rechargeable lithium batteries, J. Power Sources 138 (2004) 271–273.
[61] P. Polrolniczak, K. Wasinski, M. Walkowiak, Humic acid - derived mesoporous carbon as cathode component for lithium-sulfur battery, Int. J. Electrochem. Sci. 10 (2015) 9370–9378.
[62] P. Polrolniczak, P. Nowicki, K. Wasinski, R. Pietrzak, M. Walkowiak, Biomass-derived hierarchical carbon as sulfur cathode stabilizing agent for lithium-sulfur batteries, Solid State Ionics 297 (2016) 59–63.
[63] T.A. Zegeye, M.C. Tsai, J.H. Cheng, M.H. Lin, H.M. Chen, J. Rick, W.N. Su, C.F.J. Kuo, Controllable embedding of sulfur in high surface area nitrogen doped three dimensional reduced graphene oxide by solution drop impregnation method for high performance lithium-sulfur batteries, J. Power Sources 353 (2017) 298–311.

[64] G. Zheng, Y. Yang, J.J. Cha, S.S. Hong, Y. Cui, Hollow carbon nanofiber-encapsulated sulfur cathodes for high specific capacity rechargeable lithium batteries, Nano Lett. 11 (10) (2011) 4462–4467.
[65] L. Ji, M. Rao, S. Aloni, L. Wang, E.J. Cairns, Y. Zhang, Porous carbon nano, ber–sulfur composite electrodes for lithium/sulfur cells, Energy Environ. Sci. 4 (2011) 5053–5059.
[66] R. Elazari, G. Salitra, A. Garsuch, A. Panchenko, D. Aurbach, Sulfur-impregnated activated carbon fiber cloth as a binder-free cathode for rechargeable Li-S batteries, Adv. Mater. 23 (47) (2011) 5641–5644.
[67] J. Song, T. Xu, M.L. Gordin, P. Zhu, D. Lv, Y.B. Jiang, Y. Chen, Y. Duan, D. Wang, Nitrogen-doped mesoporous carbon promoted chemical adsorption of sulfur and fabrication of high-areal-capacity sulfur cathode with exceptional cycling stability for lithium-sulfur batteries, Adv. Funct. Mater. 24 (2014) 1243–1250.
[68] S.S. Zhang, Heteroatom-doped carbons: synthesis, chemistry and application in lithium/sulphur batteries, Inorg. Chem. Front. 2 (2015) 1059–1069.
[69] F. Chen, J. Yang, T. Bai, B. Long, X. Zhou, Biomass waste-derived honeycomb-like nitrogen and oxygen dual-doped porous carbon for high performance lithium-sulfur batteries, Electrochim. Acta 192 (2016) 99–109.
[70] C.P. Yang, Y.X. Yin, H. Ye, K.C. Jiang, J. Zhang, Y.G. Guo, Insight into the effect of boron doping on sulfur/carbon cathode in lithium–sulfur batteries, ACS Appl. Mater. Interfaces 6 (2014) 8789–8795.
[71] Y. Qiu, W. Li, W. Zhao, G. Li, Y. Hou, M. Liu, L. Zhou, F. Ye, H. Li, Z. Wei, S. Yang, W. Duan, Y. Ye, J. Guo, Y. Zhang, High-rate, ultralong cycle-life lithium/sulfur batteries enabled by nitrogen-doped graphene, Nano Lett. 14 (2014) 4821–4827.
[72] J. Yang, J. Xie, X. Zhou, Y. Zou, J. Tang, S. Wang, F. Chen, L. Wang, Functionalized N-doped porous carbon nanofiber webs for a lithium–sulfur battery with high capacity and rate performance, J. Phys. Chem. C 118 (2014) 1800–1807.
[73] J.S. Lee, A. Manthiram, Hydroxylated N-doped carbon nanotube-sulfur composites as cathodes for high-performance lithium-sulfur batteries, J. Power Sources 343 (2017) 54–59.
[74] G. Zhou, Y. Zhao, A. Manthiram, Dual-confined flexible sulfur cathodes encapsulated in nitrogen-doped double-shelled hollow carbon spheres and wrapped with graphene for Li–S batteries, Adv. Energy Mater. 5 (9) (2015) 1402263–1402273.
[75] W. Wei, J. Wang, L. Zhou, J. Yang, B. Schumann, Y. NuLi, CNT enhanced sulfur composite cathode material for high rate lithium battery, Electrochem. Commun. 13 (2011) 399–402.
[76] Y. Cao, X. Li, I.A. Aksay, J. Lemmon, Z. Nie, Z. Yang, J. Liu, Sandwich-type functionalized graphene sheet-sulfur nanocomposite for rechargeable lithium batteries, Phys. Chem. Chem. Phys. 13 (2011) 7660–7665.
[77] H.J. Peng, J.Q. Huang, M.Q. Zhao, Q. Zhang, X.B. Cheng, X.Y. Liu, W.Z. Qian, F. Wei, Nanoarchitectured graphene/CNT@Porous carbon with extraordinary electrical conductivity and interconnected micro/mesopores for lithium-sulfur batteries, Adv. Funct. Mater. 24 (2014) 2772–2781.
[78] D. Su, M. Cortie, G. Wang, Fabrication of N-doped graphene–carbon nanotube hybrids from prussian blue for lithium–sulfur batteries, Adv. Energy Mater. 7 (8) (2017) 1602014–1602026.
[79] P. Chen, T.Y. Xiao, Y.H. Qian, S.S. Li, S.H. Yu, A nitrogen-doped graphene/carbon nanotube nanocomposite with synergistically enhanced electrochemical activity, Adv. Mater. 25 (2013) 3192–3196.
[80] X. Wang, Z. Zhang, Y. Qu, Y. Lai, J. Li, Nitrogen-doped graphene/sulfur composite as cathode material for high capacity lithium–sulfur batteries, J. Power Sources 256 (2014) 361–368.

[81] S. Chen, J. Bi, Y. Zhao, L. Yang, C. Zhang, Y. Ma, Q. Wu, X. Wang, Z. Hu, Nitrogen-doped carbon nanocages as efficient metal-free electrocatalysts for oxygen reduction reaction, Adv. Mater. 24 (2012) 5593–5597.

[82] Y. Zhao, R. Nakamura, K. Kamiya, S. Nakanishi, K. Hashimoto, Nitrogen-doped carbon nanomaterials as non-metal electrocatalysts for water oxidation. Amphiphilic surface modification of hollow carbon nanofibers for improved cycle life of lithium sulfur batteries, Nano Lett. 13 (3) (2013) 1265–1270.

[83] L. Ji, M. Rao, H. Zheng, L. Zhang, Y. Li, W. Duan, J. Guo, E.J. Cairns, Y. Zhang, Graphene oxide as a sulfur immobilizer in high performance lithium/sulfur cells, J. Am. Chem. Soc. 133 (46) (2011) 18522–18525.

[84] H. Wang, Y. Yang, Y. Liang, J.T. Robinson, Y. Li, A. Jackson, Y. Cui, H. Dai, Graphene-wrapped sulfur particles as a rechargeable lithium–sulfur battery cathode material with high capacity and cycling stability, Nano Lett. 11 (7) (2011) 2644–2647.

[85] G. Zheng, Q. Zhang, J.J. Cha, Y. Yang, W. Li, Z.W. Seh, Y. Cui, Amphiphilic surface modification of hollow carbon nanofibers for improved cycle life of lithium sulfur batteries, Nano Lett. 13 (3) (2013) 1265–1270.

[86] L. Chai, J. Wang, H. Wang, L. Zhang, W. Yu, L. Mai, Porous carbonized graphene-embedded fungus film as an interlayer for superior Li–S batteries, Nano Energy 17 (2015) 224–232.

[87] L. Yin, J. Liang, G. Zhou, F. Li, R. Saito, H. Cheng, Understanding the interactions between lithium polysulfides and N-doped graphene using density functional theory calculations, Nano Energy 25 (2016) 203–210.

[88] Q. Li, M. Liu, X. Qin, J. Wu, W. Han, G. Liang, D. Zhou, Y. He, B. Li, F. Kang, Cyclized-polyacrylonitrile modified carbon nanofiber interlayers enabling strong trapping of polysulfides in lithium–sulfur batteries, J. Mater. Chem. A 4 (2016) 12973–12980.

[89] S. Niu, W. Lv, G. Zhou, H. Shi, X. Qin, C. Zheng, T. Zhou, C. Luo, Y. Deng, B. Li, Electrostatic-spraying an ultrathin, multifunctional and compact coating onto a cathode for a long-life and high-rate lithium-sulfur battery, Nano Energy 30 (2016) 138–145.

[90] X. Ji, S. Evers, R. Black, L.F. Nazar, Stabilizing lithium-sulphur cathodes using polysulphide reservoirs, Nat. Commun. 2 (2011) 325.

[91] S. Evers, T. Yim, L.F. Nazar, Understanding the nature of absorption/adsorption in nanoporous polysulfide sorbents for the Li-S battery, J. Phys. Chem. C 116 (2012) 19653–19658.

[92] R. Demir-Cakan, M. Morcrette, F. Nouar, C. Davoisne, T. Devic, D. Gonbeau, R. Dominko, C. Serre, G. Férey, J.-M. Tarascon, Cathode composites for Li-S batteries via the use of oxygenated porous architectures, J. Am. Chem. Soc. 133 (40) (2011) 16154–16160.

[93] M. Zhang, C. Yu, C. Zhao, X. Song, X. Han, S. Liu, C. Hao, J. Qiu, Cobalt-embedded nitrogen-doped hollow carbon nanorods for synergistically immobilizing the discharge products in lithium–sulfur battery, Energy Storage Mater. 5 (2016) 223–229.

[94] Q. Zhang, Y. Wang, Z.W. Seh, Z. Fu, R. Zhang, Y. Cui, Understanding the anchoring effect of two-dimensional layered materials for lithium–sulfur batteries, Nano Lett. 15 (2015) 3780–3786.

[95] X. Liang, C.Y. Kwok, F. Lodi-Marzano, Q. Pang, M. Cuisinier, H. Huang, C.J. Hart, D. Houtarde, K. Kaup, H. Sommer, Lithium-sulfur batteries: tuning transition metal oxide–sulfur interactions for long life lithium sulfur batteries: the "goldilocks" principle, Adv. Energy Mater. 6 (2016).

[96] G. Liang, J. Wu, X. Qin, M. Liu, Q. Li, Y. He, J. Kim, B. Li, F. Kang, Ultrafine TiO_2 decorated carbon nanofibers as multifunctional interlayer for high-performance lithium–sulfur battery, ACS Appl. Mater. Interfaces 8 (2016) 23105–23113.

[97] M. Liu, Q. Li, X. Qin, G. Liang, W. Han, D. Zhou, Y. He, B. Li, F. Kang, Suppressing self-discharge and shuttle effect of lithium–sulfur batteries with V_2O_5-decorated carbon nanofiber interlayer, Small (2017)1602539.
[98] X. Liu, W. Huang, D. Wang, J. Tian, Z. Shan, A nitrogen-doped 3D hierarchical carbon/sulfur composite for advanced lithium sulfur batteries, J. Power Sources 355 (2017) 211–218.
[99] S. Choudhury, B. Krüner, P. Massuti-Ballester, A. Tolosa, C. Prehal, I. Grobelsek, O. Paris, L. Borchardt, V. Presser, Microporous novolac-derived carbon beads/sulfur hybrid cathode for lithium-sulfur batteries, J. Power Sources 357 (2017) 198–208.
[100] R. Fang, S. Zhao, S. Pei, X. Qian, P. Hou, H. Cheng, C. Liu, F. Li, Toward more reliable lithium–sulfur batteries: an all-graphene cathode structure, ACS Nano 10 (2016) 8676–8682.
[101] R. Fang, S. Zhao, P. Hou, M. Cheng, S. Wang, H. Cheng, C. Liu, F. Li, 3D interconnected electrode materials with ultrahigh areal sulfur loading for Li–S batteries, Adv. Mater. 28 (2016) 3374–3382.
[102] L. Qie, A. Manthiram, A facile layer-by-layer approach for high-areal-capacity sulfur cathodes, Adv. Mater. 27 (2015) 1694–1700.
[103] L. Qie, A. Manthiram, High-energy-density lithium–sulfur batteries based on blade-cast pure sulfur electrodes, ACS Energy Lett. 1 (2016) 46–51.
[104] S. Chung, C. Chang, A. Manthiram, A core–shell electrode for dynamically and statically stable Li–S battery chemistry, Energy Environ. Sci. 9 (2016) 3188–3200.
[105] J.L. Wang, J. Yang, J.Y. Xie, N.X. Xu, A novel conductive polymer–sulfur composite cathode material for rechargeable lithium batteries, Adv. Mater. 14 (2002) 963–996.
[106] J.L. Wang, J. Yang, C.R. Wan, K. Du, J.Y. Xie, N.X. Xu, Sulfur composite cathode materials for rechargeable lithium batteries, Adv. Funct. Mater. 13 (2003) 487–492.
[107] X.U. Yu, J.Y. Xie, Y. Li, H.J. Huang, C.Y. Lai, K. Wang, Stable-cycle and high-capacity conductive sulfur-containing cathode materials for rechargeable lithium batteries, J. Power Sources 146 (2005) 335–339.
[108] X.M. He, W.H. Pu, J.U. Ren, L. Wang, J.L. Wang, C.Y. Jiang, C.R. Wan, Charge/discharge characteristics of sulfur composite electrode at different temperature and current density in rechargeable lithium batteries, Ionics 14 (2008) 335–337.
[109] X.M. He, W.H. Pu, J.G. Ren, L. Wang, J.L. Wang, C.Y. Jiang, C.R. Wan, Charge/discharge characteristics of sulfur composite cathode materials in rechargeable lithium batteries, Electrochim. Acta 52 (2007) 7372–7376.
[110] X.M. He, J.G. Ren, L. Wang, W.H. Pu, C.Y. Jiang, C.R. Wan, Expansion and shrinkage of the sulfur composite electrode in rechargeable lithium batteries, J. Power Sources 190 (2009) 154–156.
[111] X.M. He, L. Wang, W.H. Pu, J.G. Ren, W. Wu, C.Y. Jiang, C.R. Wan, Thermal analysis of sulfurization of polyacrylonitrile with elemental sulfur, J. Therm. Anal. Calorim. 94 (2008) 151–155.
[112] W.H. Pu, X.M. He, L. Wang, Z. Tian, C.Y. Jiang, C.R. Wan, Application of triethylammonium salts as ionic liquid catalyst and medium for Fischer esterification, Ionics 13 (2007) 273–276.
[113] J. Wang, J. Chen, K. Konstantinov, L. Zhao, S.H. Ng, G.X. Wang, Z.P. Guo, H.K. Liu, Sulphur-polypyrrole composite positive electrode materials for rechargeable lithium batteries, Electrochim. Acta 51 (2006) 4634–4638.
[114] X. Liang, Z. Wen, Y. Liu, X. Wang, H. Zhang, M. Wu, L. Huang, Preparation and characterization of sulfur–polypyrrole composites with controlled morphology as high capacity cathode for lithium batteries, Solid State Ionics 192 (2011) 347–350.
[115] M. Sun, S. Zhang, T. Jiang, L. Zhang, J. Yu, Nano-wire networks of sulfur–polypyrrole composite cathode materials for rechargeable lithium batteries, Electrochem. Commun. 10 (2008) 1819–1822.

[116] L. Qiu, S. Zhang, L. Zhang, M. Sun, W. Wang, Preparation and enhanced electrochemical properties of nano-sulfur/poly(pyrrole-co-aniline) cathode material for lithium/sulfur batteries, Electrochim. Acta 55 (2010) 4632–4636.
[117] F. Wu, S.X. Wu, R.J. Chen, J.Z. Chen, S. Chen, Sulfur–polythiophene composite cathode materials for rechargeable lithium batteries, Electrochem. Solid-State Lett. 13 (2010) A29–A31.
[118] W. Li, Q. Zhang, G. Zheng, Z.W. Seh, H. Yao, Y. Cui, Understanding the role of different conductive polymers in improving the nanostructured sulfur cathode performance, Nano Lett. 13 (11) (2013) 5534–5540.
[119] T. Sotomura, H. Uemachi, K. Takeyama, New organodisulfide-polyaniline composite cathode for secondary lithium battery, Electrochim. Acta 37 (1992) 1851–1854.
[120] P. Novak, K. Muller, K.S.V. Santhanam, O. Haas, Electrochemically active polymers for rechargeable batteries, Chem. Rev. 97 (1997) 207–281.
[121] M. Liu, S.J. Visco, L.C. De Jonghe, Novel solid redox polymerization electrodes. all-solid-state, thin-film, rechargeable lithium batteries, J. Electrochem. Soc. 138 (1991) 1891–1895.
[122] S.J. Visco, M. Liu, L.C. De Jonghe, Ambient temperature high-rate lithium/organosulfur batteries, J. Electrochem. Soc. 137 (1990) 1191–1192.
[123] L.J. Xue, J.X. Li, S.Q. Hu, M.X. Zhang, Y.H. Zhou, C.M. Zhan, Anthracene based organodisulfide positive active materials for lithium secondary battery, Electrochem. Commun. 5 (2003) 903–906.
[124] J. Li, H. Zhan, L. Zhou, S. Deng, Z. Li, Y. Zhou, Aniline-based polyorganodisulfide redox system of high energy for secondary lithium batteries, Electrochem. Commun. 6 (2004) 515–519.
[125] S.C. Zhang, L. Zhang, W.K. Wang, W.J. Xue, A novel cathode material based on polyaniline used for lithium/sulfur secondary battery, Synth. Met. 160 (2010) 2041–2044.
[126] M.M. Doeff, M.M. Lerner, S.J. Visco, L.C. De Jonghe, The use of polydisulfides and copolymeric disulfides in the Li/PEO/SRPE battery system, J. Electrochem. Soc. 139 (1992) 2077–2081.
[127] N. Oyama, T. Tatsuma, T. Sato, T. Sotomura, Dimercaptan–polyaniline composite electrodes for lithium batteries with high energy density, Nature 373 (1995) 598–600.
[128] B.A. Trofimov, A.G. Mal'kina, I.A. Dorofeev, G.F. Myachina, I.V. Rodionova, T.I. Vakul'skaya, L.M. Sinegovskaya, T.A. Skotheim, Synthesis and properties of polyenic oligosulfides derived from acetylene and elemental sulfur, Russ. J. Gen. Chem. 77 (2007) 1559–1566.
[129] B.A. Trofimov, A.G. Mal'kina, I. Dorofeev, G.F. Myachina, I.V. Rodionova, T.I. Vakul'skaya, L.M. Sinegovskaya, T.A. Skotheim, Ethynedithiol oligomers as cathode components of lithium-sulfur batteries, Dokl. Chem. 414 (2007) 125–127.
[130] B.A. Trofimov, L.V. Morozova, M.V. Markova, A.I. Mikhaleva, G.F. Myachina, I.V. Tatarinova, T.A. Skotheim, Vinyl ethers with polysulfide and hydroxyl functions and polymers therefrom as binders for lithium–sulfur batteries, J. Appl. Polym. Sci. 101 (2006) 4051–4055.
[131] B.A. Trofimov, G.F. Myachina, I.V. Rodionova, A.G. Mal'kina, I.A. Dorofeev, T.I. Vakul'skaya, L.M. Sinegovskaya, T.A. Skotheim, Ethynedithiol-based polyeneoligosulfides as active cathode materials for lithium-sulfur batteries, J. Appl. Polym. Sci. 107 (2008) 784–787.
[132] X.G. Yu, J.Y. Xie, J. Yang, H.J. Huang, K. Wang, Z.S. Wen, Lithium storage in conductive sulfur-containing polymers, J. Electroanal. Chem. 573 (2004) 121–128.
[133] S.-H. Chung, C.-H. Chang, A. Manthiram, Development of high-performance, high-loading sulfur cathodes, ECS MA2016-03 1139.

[134] R.V. Bugga, S.C. Jones, J. Pasalic, C.S. Seu, J.-P. Jones, L. Torres, Metal sulfide-blended sulfur cathodes in high energy lithium-sulfur cells, J. Electrochem. Soc. 164 (2) (2017) A265–A276.
[135] F. Pei, L. Lin, D. Ou, Z. Zheng, S. Mo, X. Fang, N. Zheng, Self-supporting sulfur cathodes enabled by two-dimensional carbon yolk-shell nanosheets for high-energy-density lithium-sulfur batteries. Nat. Commun. 8 (2017) 182, https://doi.org/10.1038/s41467-017-00575-8.
[136] C.-H. Chang, S.-H. Chung, A. Manthiram, Effective stabilization of a high-loading sulfur cathode and a lithium-metal anode in Li-S batteries utilizing SWCNT-modulated separators, Small 12 (2) (2016) 174–179.
[137] L. Luo, A. Manthiram, Rational design of high-loading sulfur cathodes with a poached-egg-shaped architecture for long-cycle lithium–sulfur batteries, ACS Energy Lett. 2 (10) (2017) 2205–2211.
[138] X. Wang, T. Gao, F. Han, Z. Ma, Z. Zhang, J. Li, C. Wang, Stabilizing high sulfur loading Li–S batteries by chemisorption of polysulfide on three-dimensional current collector, Nano Energy 30 (2016) 700–708.
[139] F. Qin, X. Wang, K. Zhang, J. Fang, J. Li, Y. Lai, High areal capacity cathode and electrolyte reservoir render practical Li-S batteries, Nano Energy 38 (2017) 137–146.
[140] H. Xu, A. Manthiram, Hollow cobalt sulfide polyhedra-enabled long-life, high areal-capacity lithium-sulfur batteries, Nano Energy 33 (2017) 124–129.
[141] E.C. Gay, D.R. Vissers, F.J. Martino, K.E. Anderson, Performance characteristics of solid lithium-aluminum alloy electrodes, J. Electrochem. Soc. 123 (1976) 1591–1596.
[142] D. Golodnitsky, E. Peled, Pyrite as cathode insertion material in rechargeable lithium/composite polymer electrolyte batteries, Electrochim. Acta 45 (1999) 335–350.
[143] E. Strauss, D. Golodnitsky, E. Peled, Study of phase changes during 500 full cycles of Li/composite polymer electrolyte/FeS_2 battery, Electrochim. Acta 45 (2000) 1519–1525.
[144] Y.J. Choi, J.H. Jeong, J.H. Ha, G.B. Cho, K.K. Cho, K.S. Ryu, K.W. Kim, Electrochemical properties of Li–Fe–S ternary metal sulfide (lithium iron sulfide) synthesized via the molten salt method, Phys. Scr. T139 (2010) 14063.
[145] M.N. Obrovac, J.R. Dahn, Electrochemically active lithia/metal and lithium sulfide/metal composites, Electrochem. Solid-State Lett. 5 (2002) A70–A73.
[146] Y.N. Zhou, C.L. Wu, H. Zhang, X.J. Wu, Z.W. Fu, Electrochemical reactivity of Co–Li_2S nanocomposite for lithium-ion batteries, Electrochim. Acta 52 (2007) 3130–3136.
[147] A. Hayashi, T. Ohtomo, F. Mizuno, K. Tadanaga, M. Tatsumisago, All-solid-state Li/S batteries with highly conductive glass–ceramic electrolytes, Electrochem. Commun. 5 (2003) 701–705.
[148] A. Hayashi, R. Ohtsubo, T. Ohtomo, F. Mizuno, M. Tatsumisago, All-solid-state rechargeable lithium batteries with Li_2S as a positive electrode material, J. Power Sources 183 (2008) 422–426.
[149] B.M.L. Rao, J.A. Shropshire, Effect of sulfur impurities on Li / TiS2 cells, J. Electrochem. Soc. 128 (1981) 942–945.
[150] J.H. Kwon, H.J. Ahn, M.S. Jeon, K.W. Kim, I.S. Ahn, J.H. Ahn, G.X. Wang, H.S. Ryu, The electrochemical properties of Li/TEGDME/MoS_2 cells using multi-wall carbon nanotubes as a conducting agent, Res. Chem. Intermed. 36 (2010) 749–759.
[151] C.H. Lai, K.W. Huang, J.H. Cheng, C.Y. Lee, B.J. Hwang, L.J. Chen, Direct growth of high-rate capability and high capacity copper sulfide nanowire array cathodes for lithium-ion batteries, J. Mater. Chem. 20 (2010) 6638–6645.
[152] C.H. Lai, K.W. Huang, J.H. Cheng, C.Y. Lee, W.F. Lee, C.T. Huang, B.J. Hwang, L.J. Chen, Oriented growth of large-scale nickel sulfide nanowire arrays via a general

solution route for lithium-ion battery cathode applications, J. Mater. Chem. 19 (2009) 7277–7283.
[153] Z. Wang, L. Pan, H. Hu, S. Zhao, Co_9S_8 nanotubes synthesized on the basis of nanoscale Kirkendall effect and their magnetic and electrochemical properties, CrystEngComm 12 (2010) 1899–1904.
[154] C. Zhai, N. Du, H.Z.D. Yang, Large-scale synthesis of ultrathin hexagonal tin disulfide nanosheets with highly reversible lithium storage, Chem. Commun. 47 (2011) 1270–1272.
[155] G.X. Xu, Q. Lu, B.T. Yu, L. Wen, Inorganic polymer phosphazene disulfide as cathode material for rechargeable lithium batteries, Solid State Ionics 177 (2006) 305–309.
[156] Z. Jiao, L. Chen, J. Si, C. Xu, Y. Jiang, Y. Zhu, Y. Yang, B. Zhao, Core-shell $Li_2S@Li_3PS_4$ nanoparticles incorporated into graphene aerogel for lithium-sulfur batteries with low potential barrier and overpotential, J. Power Sources 353 (2017) 167–175.
[157] L. Jin, G. Li, B. Liu, Z. Li, J. Zheng, J.P. Zheng, A novel strategy for high-stability lithium sulfur batteries by in situ formation of polysulfide adsorptive-blocking layer, J. Power Sources 355 (2017) 147–153.
[158] A. Konarov, Z. Bakenov, H. Yashiro, Y.K. Sun, S.T. Myung, Effect of carbon-sulphur bond in a sulphur/dehydrogenated polyacrylonitrile/reduced graphene oxide composite cathode for lithium-sulphur batteries, J. Power Sources 355 (2017) 140–146.
[159] X. Huang, K. Shi, J. Yang, G. Mao, J. Chen, MnO_2-GO double-shelled sulfur (S@MnO_2@GO) as a cathode for Li-S batteries with improved rate capability and cyclic performance, J. Power Sources 356 (2017) 72–79.
[160] X. Song, T. Gao, S. Wang, Y. Bao, G. Chen, L.X. Ding, H. Wang, Free-standing sulfur host based on titanium-dioxide-modified porouscarbon nanofibers for lithium-sulfur batteries, J. Power Sources 356 (2017) 172–180.
[161] S.Z. Zeng, Y. Yao, X. Zeng, Q. He, X. Zheng, S. Chen, W. Tu, J. Zou, A composite of hollow carbon nanospheres and sulfur-rich polymers for lithium-sulfur batteries, J. Power Sources 357 (2017) 11–18.
[162] Y. Yang, G. Zheng, Y. Cui, A membrane-free lithium/polysulfide semi-liquid battery for large-scale energy storage, Energy Environ. Sci. 6 (2013) 1552–1558.
[163] H. Pan, X. Wei, W.A. Henderson, Y. Shao, J. Chen, P. Bhattacharya, J. Xiao, J. Liu, On the way toward understanding solution chemistry of lithium polysulfides for high energy Li–S redox flow batteries, Adv. Energy Mater. 5 (2015).
[164] G. Zhou, Y. Zhao, C. Zu, A. Manthiram, Free-standing TiO2 nanowire-embedded graphene hybrid membrane for advanced Li/dissolved polysulfide batteries, Nano Energy 12 (2015) 240–249.
[165] H. Chen, Q. Zou, Z. Liang, H. Liu, Q. Li, Y. Lu, Sulphur-impregnated flow cathode to enable high-energy-density lithium flow batteries, Nat. Commun. 6 (2015) 5877.
[166] F.Y. Fan, W.H. Woodford, Z. Li, N. Baram, K.C. Smith, A. Helal, G.H. McKinley, W.C. Carter, Y. Chiang, Polysulfide flow batteries enabled by percolating nanoscale conductor networks, Nano Lett. 14 (2014) 2210–2218.
[167] Z. Liang, G. Zheng, W. Li, Z.W. Seh, H. Yao, K. Yan, D. Kong, Y. Cui, Sulfur cathodes with hydrogen reduced titanium dioxide inverse opal structure, ACS Nano 8 (5) (2014) 5249–5256.
[168] X. Tao, J. Wang, Z. Ying, Q. Cai, G. Zheng, Y. Gan, H. Huang, Y. Xia, C. Liang, W. Zhang, Y. Cui, Strong sulfur binding with conducting magnéli-phase TinO2n–1 nanomaterials for improving lithium–sulfur batteries, Nano Lett. 14 (9) (2014) 5288–5294.
[169] H. Wang, Q. Zhang, H. Yao, Z. Liang, H.-W. Lee, P.-C. Hsu, G. Zheng, Y. Cui, High electrochemical selectivity of edge versus terrace sites in two-dimensional layered MoS2 materials, Nano Lett. 14 (12) (2014) 7138–7144.

CHAPTER SIX

Binders

The binder is an essential component because it plays an important role in maintaining the structural integrity of the electrodes, and thus the capacity and the ability to operate cyclic lithium-sulfur batteries. A good binder should create a good and stable network between sulfur and conductive materials, facilitating electronic and ionic transport [1].

According to [2], binders should aid in electrode processing and drying on to aluminum current collectors, electrolyte wetting during cell assembly, ion transport, and mechanical integrity upon cycling to accommodate the volume changes associated with S_8-Li_2S interconversion.

Several alternative electrode binders, such as polyethylene oxide (PEO), gelatin, polyvinyl pyrrolidone (PVP), Na-alginate, gum arabic (GA), etc., have been studied as alternatives to common polyvinylidene difluoride (PVDF) [3–8]. According to [2], also PVP blends with Nafion, PAMAM dendrimers, polycationic-cyclodextrins, poly(acrylic acid), poly-(ethylene oxide), and carboxymethyl-cellulose:styrenebutadiene-rubber (CMC: SBR), have been investigated as the binders aimed to improve cathode performance. Some binders mitigated the migration of soluble polysulfides from the cathode into the electrolyte, which otherwise would lead to stranded sulfur in the cell or instabilities in the lithium anode. None has been reported that directly participated in the redox chemistry of sulfur or otherwise served to enhance ion transport as needed for high-rate applications. The ideal binder is needed to create more robust conductive structures and improve the strength and structural stability of the entire sulfur electrode.

According to [9], capacity fading on cycling of lithium-sulfur batteries results from at least four processes: increase of SEI thickness resistance, loss of cathode capacity (precipitation of sulfur species outside the cathode), agglomeration and thickening of sulfur species, and increase in cell impedance because of reduction of the electrolyte. An important issue that has not been properly addressed up to now is the influence of the type and content of the cathode binder on the cell parameters and on the electrochemical performance of lithium-sulfur batteries.

Next-generation Batteries with Sulfur Cathodes
https://doi.org/10.1016/B978-0-12-816392-4.00006-2

PEO shows good ionic conductivity resulting from its coordination with lithium ions [10]. Using PEO as a binder in the cathode, 80% of the theoretical capacity is provided for the first discharge, and the capacity of 100 $mAh\,g^{-1}$ is maintained at a temperature of 80°C. The morphology and electrochemical properties of the cathode depend on the processing procedure and the type of binder: a mechanically mixed mixture of PEO, carbon, and sulfur forms a porous PEO film that has better cyclic performance than the milled ball mill mixture that forms a dense and thick PEO layer over the sulfur particles. In contrast, when poly(butadiene-*co*-styrene) (PBS) serves as a tackifier in a ball mill, the dry mixture has a good ability to work cyclically [11]. In general, PEO exhibits poor adhesive properties [12, 13] and low temperature ionic conductivity [14], and electrodes made from it show better performance at elevated temperatures. The use of polytetrafluoroethylene (PTFE)-carboxymethyl cellulose (CMC) composites as binders improves the capacity and use of sulfur in comparison with polyvinyl pyrrolidone (PVP) and PTFE-polyvinyl alcohol (PVA) composite used as binders [15], but no cyclic performance is reported. In another work, the capacity increases, as noted, along with the number of cycles up to the 10th cycle, using a PTFE-CMC composite as a binder [16]. According to [17], the discharging capacity and cyclic work efficiency have been improved using a combination of PVP and polyethylene imine (PEI) compared to a single component PVP binder. The chemical reaction between PVP and PEI increases the stability of the PVP-PEI composite in the electrolyte, giving good mechanical strength to the cathode. Gelatin is a recently developed binder for a sulfur electrode.

The cathodes utilizing gelatin as a binder have a better initial capacity and cyclic capacity than those using PEO as a binder [1, 18–20]. Gelatin exhibits favorable properties of high stability in electrolytes, effective adhesion, and dispersion of positive electrode materials.

In [7], a binder inspired by Acacia Senegal, GA, is presented, which is nontoxic and balanced for high-performance lithium-sulfur batteries. The GA structure is shown in Fig. 22.

Fig. 23A shows the discharging/loading profiles of the link with the binder GA at C/5. It was evident that there are two typical plateaus during discharge, corresponding to the form at ion of soluble long-chain polysulfides and insoluble short-chain Li_2S_2/Li_2S. As shown in Fig. 23B, S@GA could provide an initial capacity of 1386 $mAh\,g^{-1}$, which was significantly higher than that of S@Gelatin and S@PVDF. A large 1090 $mAh\,g^{-1}$ capacity could still be reached after 50 cycles for S@GA, indicating the excellent

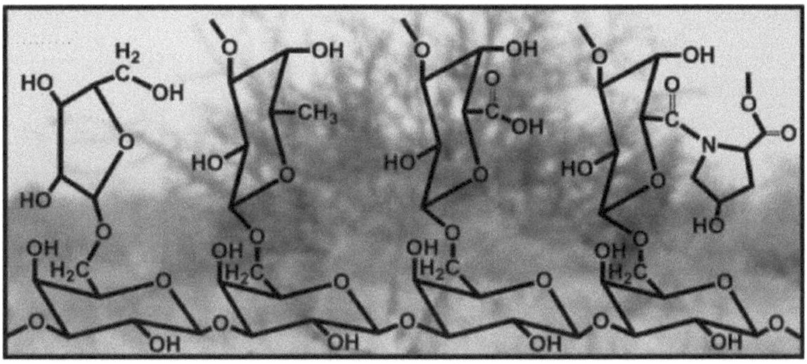

Fig. 22 A schematic image of the GA chemical structure with Acacia Senegal background [7].

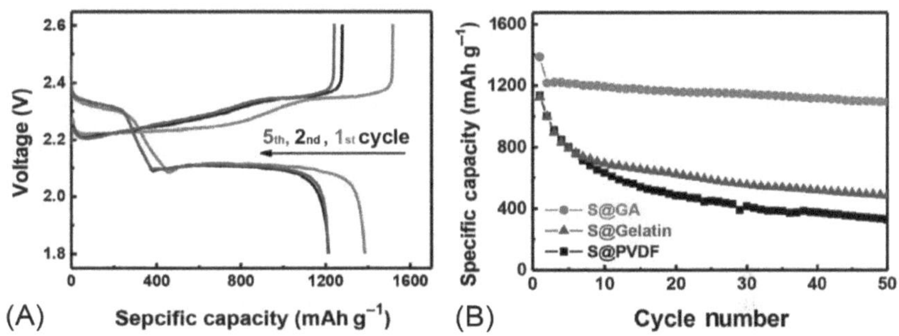

Fig. 23 (A) Cell discharging/loading profiles with GA binder at C/5; (B) the performance of cyclic work of cells with different binders [7].

integrity of the cell electrode from the GA. On the other hand, cell capacities with gelatin and PVDF decreased to 452 and 312 $mAh\,g^{-1}$, respectively.

Although these alternative positive binders affect lithium-sulfur batteries, there are still many limitations to their practical applications due to the unclear way these binders interact with the sulfur cathode to increase performance. Therefore, more research needs to be done to understand this mechanism.

Some other modern binders are characterized in the following paragraphs.

In [9] the electrochemical behavior, during prolonged cycling, of Li_2S-based cathodes containing five different binders was analyzed. These binders were: poly(vinylidene fluoride) (PVDF-HFP), polyvinyl pyrrolidone (PVP), mix of PVP with polyethyleneimine (PEI), polyaniline (PANI), and lithium polyacrylate (LiPAA). Sulfur utilization in the cathode followed the order of LiPAA > PVP:PEI > PVP > PVDF-HFP > PANI. Depending on the type of binder, cells provided 500–1400 mAh $g(S)^{-1}$, 94.6%–98.0% faradaic efficiency, and enabled more than 500 reversible cycles.

In [21] a poly(acrylamide-*co*-diallyl dimethyl ammonium chloride) (AMAC) was introduced as the binder of high sulfur loading cathodes. AMAC, a cationic polyelectrolyte binder, was substantially insoluble in organic solvent while being highly soluble in water. Such features enabled retaining the stable void structure of the sulfur cathode. However, the ionic conductivity of the AMAC and sulfur aqueous slurry caused severe pitting corrosion of the Al current collector due to the formation of galvanic cells between sulfur particles and Al metal in the process of aqueous slurry coating. A dual-binder approach allowed effective elimination of the Al corrosion while still retaining the advantage of the AMAC binder.

In [22], a redox-active supramolecular polymer binder was elaborated. It utilized a perylenebisimide (PBI) scaffold featuring four carboxylic acid moieties that can be lithiated by treatment with LiOH. Upon lithiation, the binder was water soluble, which permits aqueous processing, thereby maximizing the environmental and cost benefits of Li-S cells. Sulfur slurries containing the water soluble, lithiated PBI polymer binder were casted onto an Al foil current collector. Once casted, the binders self-assembled through -stacking interactions of the aromatic cores into nanowire architectures.

The obtained nanowire web morphology provided stronger physical binding between active S particles and the current collector. This avoided electronic disconnection between the electrode components, resulting in longer Li-S cell lifetimes. The lithiated nanowire structured electrodes demonstrated a 58% increase in capacity retention after 250 cycles at 1.5 C compared to nonlithiated control samples without modified nanowire architecture.

In [2], -stacked perylenebisimide (PBI) molecules were utilized as highly networked, redox-active supramolecular polymer binders in sulfur cathodes for lightweight and energy-dense Li-S batteries. The in-operando reduction and lithiation of these PBI binders sustainably reduced Li-S cell impedance relative to nonredox active conventional polymer binders. That lower impedance enabled high-rate cycling in Li-S cells with excellent durability.

Paper [23] presented a multifunctional polar binder (AHP) by polymerization of hexamethylene diisocyanate (HDI) with ethylenediamine (EDA) bearing many amino groups, which were used in electrode preparation with commercial sulfur powder cathodes. The abundant amide groups of the binder endowed the cathode with multidimensional chemical bonding interaction with sulfur species within the cathode to inhibit the shuttling effect of polysulfides, while the suitable ductility to buffer volume change. Utilizing such features, composite C/S cathodes based on the binder displayed excellent capacity retention at 0.5 C, 1 C, 1.5 C, and 3 C over 200 cycles.

In [24], the molecular-level underpinnings distinguishing an active polyelectrolyte binder designed for lithium-sulfur batteries from a passive alternative were revealed. The binder, a cationic polyelectrolyte, was shown to both facilitate lithium-ion transport through its reconfigurable network of mobile anions and restrict polysulfide diffusion from mesoporous carbon hosts by anion metathesis, which is selective for higher oligomers. Such attributes allowed cells operation for >100 cycles with excellent rate capability using cathodes with real sulfur loadings up to $8.1\,mg\,cm^{-2}$.

According to [25], a sulfonated polystyrene (SPS) prepared via homogeneous reaction was used as a functional binder for the sulfur cathode of lithium sulfur cells. The traditional binder polyvinylidene fluoride (PVDF) was used for comparison. For the sulfur cathode with PVDF as binder, the capacity retention after 100 cycles at $200\,mA\,g^{-1}$ was 46.9% and severe voltage fading performance from the 10th to 100th cycle occurred. For the SPS binder, the capacity retention after 100 cycles was 74.4% and there was almost no change of the first plateau at around 2.3 V in the discharge curve from the 10th to 100th cycle, indicating electrochemical performance improvement of the lithium-sulfur cell.

REFERENCES

[1] J. Sun, Y.Q. Huang, W.K. Wang, Z.B. Yu, A.B. Wang, K.G. Yuan, Electrochim. Acta 53 (2008) 7084–7088.

[2] P.D. Frischmann, Y. Hwa, E.J. Cairns, B.A. Helms, Redox-active supramolecular polymer binders for lithium sulfur batteries that adapt their transport properties in operando, Chem. Mater. 28 (2016) 7414–7421, https://doi.org/10.1021/acs.chemmater.6b03013.

[3] X. Duan, Y. Han, Y. Li, Y. Chen, Improved capacity retention of low cost sulfur cathodes enabled by a novel starch binder derived from food, RSC Adv. 4 (105) (2014) 60995–61000.

[4] M.J. Lacey, F. Jeschull, K. Edström, D. Brandell, Why PEO as a binder or polymer coating increases capacity in the Li–S system, Chem. Commun. 49 (76) (2013) 8531–8533.

[5] J. Pan, G. Xu, B. Ding, J. Han, H. Dou, X. Zhang, Enhanced electrochemical performance of sulfur cathodes with a water-soluble binder, RSC Adv. 5 (18) (2015) 13709–13714.

[6] Y. Chen, N. Liu, H. Shao, W. Wang, M. Gao, C. Li, H. Zhang, A. Wang, Y. Huang, Chitosan as a functional additive for high-performance lithium–sulfur batteries, J. Mater. Chem. A 3 (29) (2015) 15235–15240.

[7] G. Li, M. Ling, Y. Ye, Z. Li, J. Guo, Y. Yao, J. Zhu, Z. Lin, S. Zhang, Acacia senegal–inspired bifunctional binder for longevity of lithium–sulfur batteries, Adv. Energy Mater. 5 (21) (2015).
[8] G. Li, W. Cai, B. Liu, Z. Li, A multi functional binder with lithium ion conductive polymer and polysulfide absorbents to improve cycleability of lithium–sulfur batteries, J. Power Sources 294 (2015) 187–192.
[9] E. Peled, M. Goor, I. Schektman, T. Mukra, Y. Shoval, D. Golodnitsky, The effect of binders on the performance and defradation of the lithium/sulfur battery assembled in the discharged state, J. Electrochem. Soc. 164 (1) (2017) A5001–A5007.
[10] J.H. Shin, K.W. Kim, H.J. Ahn, J.H. Ahn, Mater. Sci. Eng. B Solid State Mater. Adv. Technol. 95 (2002) 148–156.
[11] S.-E. Cheon, S.-S. Choi, J.-S. Han, Y.-S. Choi, B.-H. Jung, H.S. Lim, J. Electrochem. Soc. 151 (2004) A2067–A2073.
[12] S.E. Cheon, K.S. Ko, J.H. Cho, S.W. Kim, E.Y. Chin, H.T. Kim, J. Electrochem. Soc. 150 (2003) A800–A805.
[13] S.-E. Cheon, J.-H. Cho, K.-S. Ko, C.-W. Kwon, D.-R. Chang, H.-T. Kim, S.-W. Kim, J. Electrochem. Soc. 149 (2002) A1437–A1441.
[14] F. Croce, L. Persi, F. Ronci, B. Scrosati, Solid State Ionics 135 (2000) 47–52.
[15] N.-I. Kim, C.-B. Lee, J.-M. Seo, W.-J. Lee, Y.-B. Roh, J. Power Sources 132 (2004) 209–212.
[16] H.-S. Ryu, H.-J. Ahn, K.-W. Kim, J.-H. Ahn, K.-K. Cho, T.-H. Nam, J.-U. Kim, G.-B. Cho, J. Power Sources 163 (2006) 201–206.
[17] Y.J. Jung, S. Kim, Electrochem. Commun. 9 (2007) 249–254.
[18] W.K. Wang, Y. Wang, Y.Q. Huang, C.J. Huang, Z.B. Yu, H. Zhang, A.B. Wang, K.G. Yuan, J. Appl. Electrochem. 40 (2010) 321–325.
[19] J. Sun, Y.Q. Huang, W.K. Wang, Z.B. Yu, A.B. Wang, K.G. Yuan, Electrochem. Commun. 10 (2008) 930–933.
[20] W.Y. Zhang, Y.Q. Huang, W.K. Wang, C.J. Huang, Y. Wang, Z.B. Yu, H. Zhang, J. Electrochem. Soc. 157 (2010) A443–A446.
[21] S.S. Zhang, Binder based on polyelectrolyte for high capacity density lithium/sulfur battery, J. Electrochem. Soc. 159 (8) (2012) A1226–A1229, https://doi.org/10.1149/2.039208jes.
[22] Y. Hwa, P.D. Frischmann, B.A. Helms, E.J. Cairns, Aqueous-processable redox-active supramolecular polymer binders for advanced lithium/sulfur cells, Chem. Mater. 30 (3) (2018) 685–691, https://doi.org/10.1021/acs.chemmater.7b03870.
[23] Y. Jiao, W. Chen, T. Lei, L. Dai, B. Chen, C. Wu, J. Xiong, A novel polar copolymer design as a multi-functional binder for strong affinity of polysulfides in lithium-sulfur batteries, Nanoscale Res. Lett. 12 (2017) 195, https://doi.org/10.1186/s11671-017-1948-5.
[24] L. Li, T.A. Pascal, J.G. Connell, F.Y. Fan, S.M. Meckler, L. Ma, Y.-M. Chiang, D. Prendergast, B.A. Helms, Molecular understanding of polyelectrolyte binders that actively regulate ion transport in sulfur cathodes. Nat. Commun. 8 (2017)https://doi.org/10.1038/s41467-017-02410-6. Article number: 2277.
[25] M. Cheng, Y. Liu, X. Guo, Z. Wu, y. Chen, J. Li, L. Li, B. Zhong, A novel binder-sulfonated polystyrene for the sulfur cathode of Li-S batteries, Ionics 23 (2017) 2251–2258, https://doi.org/10.1007/s11581-017-2087-9.

CHAPTER SEVEN

Electrolytes

The electrolyte is another vital component of lithium-sulfur (Li-S) batteries, and the selection of each electrolyte component directly affects the battery performance [1–3]. Electrolytes provide ion transport between cathode and anode and play a key role in lithium batteries. Thanks to their low operating potential (>3.6 V), Li-S batteries have no problems related to the decomposition of liquid electrolytes, which is observed in high-potential lithium-ion batteries. Nevertheless, Li-S batteries in which liquid electrolytes are used experience a dramatic loss of capacity, due to the high solubility of polysulfides in liquid electrolytes. Polysulfide transfer, corrosion of the lithium anode, and irreversible deposition of less soluble Li_2S_2 and Li_2S are associated with high solubility of polysulfides, which results in high mobility of sulfur in the electrochemical cell. Many problems in Li-S batteries can be solved with solid electrolytes, such as gel and ceramic electrolytes, although solid electrolytes have worse ionic conductivity and slower kinetics.

7.1 LIQUID ELECTROLYTES

Liquid electrolytes include lithium salts dissolved in organic solvents. It is widely accepted that the electrolyte composition has a considerable influence on the efficiency of charging and discharging the negative lithium electrode in lithium-ion batteries [4]. In lithium-sulfur batteries, the problems are not only related to the lithium anode, but also to the sulfur cathode [5]. In addition to the corrosion and passivation of the lithium anode by soluble polysulfide types, the liquid electrolyte must meet the following requirements:

(i) low viscosity,
(ii) high polysulfide solubility,
(iii) good electrochemical properties in the sulfur redox reaction.

For example, sulfur can only be reduced to S2-4 in dimethyl sulfoxide (DMSO), which corresponds to 25% of the maximum sulfur use. The maximum use of sulfur is 50% in dimethylacetamide (DMAC) because sulfur can be reduced to S2-2. However, in THF, about 100% of the use of sulfur is obtained at 50°C [6]. The nature of organic solvents and lithium salts has a

https://doi.org/10.1016/B978-0-12-816392-4.00007-4

considerable influence on electrochemical processes involving sulfur compounds on the positive electrode [7, 8]. Decision making in electrolyte selection is crucial for the use of active materials and cyclic Li-S batteries.

As stated in [9], liquid electrolytes are often used, not only in Li-S batteries, but also in most other electrochemical cells, due to low surface tension and low viscosity. These properties ensure good electrolyte wetting, which determines a good contact between the electrolyte and active materials, with low contact resistance [10]. At the same time, they facilitate the use of electrolytes with high ionic conductivity, because of facilitating ion diffusion and the formation of a homogeneous solution. In modern Li-S cells, the electrolyte must be subject to specific requirements, due to the high chemical reactivity of PS and metallic lithium [11, 12]. In addition, considering the weak electrochemical kinetics of sulfur and lithium sulfide, more advantageous realizations of electronic and ionic paths make optimal selection of the Li-S cell electrolyte more difficult [10].

7.1.1 Ether-Based Electrolytes

Currently, the ether-based electrolyte is the most commonly used electrolyte in Li-S batteries, usually consisting of 1 M amidium bis-trifluoromethanesulfonimide (LiTFSI) in a binary ether solvent 1,2-dimethoxyethane (DME) and 1,3-dioxolane (DOL) with the addition of a small amount of $LiNO_3$ [11–14]. Ether solvents are chosen due to their general stability in the presence of PS [15–17] and high ion conductivity and better electrolyte/electrode contact resulting from low viscosity and perfect wettability [12].

7.1.2 Solvents

Various electrolytes have been investigated so far, including one-component, two-component, and even ternary solvents with appropriate lithium salts. One-component solvents include THF [18–20], DME [21], 1,3-dioxolane (DOL) [22], tri (ethylene glycol) dimethyl ether (triglyme) [23], tetra (glycol) ethylene dimethyl ether (TEGDME) [22–24], and polyethylene glycol dimethyl ether (PEGDME) [23]. Usually a one-component solvent does not meet all three electrolyte conditions [22].

The high viscosity of the concentrated polysulfide solution in THF leads to a poor use of sulfur. A small portion of toluene in THF significantly reduces the viscosity of the polysulfide solution in THF, thus improving the use of sulfur above 95% [5], but the capacity increase is occupied by low maximum efficiency, due to the low conductivity of the THF-toluene solution. The addition of DOL to THF improves conductivity, but the use of sulfur is only 50% due to the incomplete discharge of sulfur; the discharging product is limited to Li2S2 in a DOL-rich electrolyte [25].

Among the many types of ether solvents, a chain ether such as DME, polyethylene glycol dimethyl ether (PEGDME), and ethylene glycol dimethyl ether (TEGDME) are most often used as a single solvent in Li-S batteries because of their good PS dissolution capacity [19, 26–28]. Table 2 shows the list of solvents used in the current Li-S cells and their properties. For example, [32] describes the study of a single DME, a single DOL and a DME + DOL binary solvent as an electrolyte for a Li-S battery. DOL had a relatively lower solubility of the polysulfide but provided a more stable solid electrolyte (SEI) contact on the surface of the metallic lithium. DME, on the other hand, is more reactive with lithium, but provides better solubility of the polysulfide to improve cathode reaction kinetics. These two electrolyte solvents show some synergistic effect to improve the proper capacity and lifetime of lithium cells.

A two-component mixture of DOL and DME has an excellent ability to work cyclically [33]. Polymerization of DOL partially protects the lithium anode.

During the tests presented in [34], different compositions of DME electrolyte were used: DOL = 4:1, 2:1, 1:1, 1:2 and 1:4 as solvent and $LiClO_4$ as the lithium salt. For the optimal DME composition: DOL = 2:1 (v/v), the cell exhibited the highest initial capacity of 1200 mAh g^{-1} and the best maintained cyclic capacity of 750 mAh g^{-1} after 20 cycles. The EIS analysis applied later indicated that the high content of DME in the electrolyte

Table 2 List of Solvents Used in Current Li-S Cells and Their Properties [28–31]

Solvent		M_w	T_b (°C)	T_m (°C)	μ (cP)	ε
DME	1,2-Dimethoxyethane	90.12	85	−69	0.46	7.3
TEGDME	Tetraethylene glycol dimethyl ether	222.28	275	−30	4.05	7.9
DOL	1,3-Dioxolane	74.08	78	−95	0.6	7.1
EEE	2-Ethoxy ethyl ether	162.63	188	−44.3	1.4	–
PEGDME	Polyethylene glycol dimethyl ether	~250	>250	−23	7.0	–
Triglyme	Triethylene glycol dimethyl ether	178.23	216	−45	1.95	–
DEGDME	Diethylene glycol dimethyl ether	134.18	162	−68	0.99	7.3
THF	Tetrahydrofuran	72.1	166	−108.4	0.48	7.6

M_w, molecular weight; T_b, normal boiling point; T_m, normal melting point; μ, viscosity at 25°C; ε, dielectric constant.

increased the contact resistance due to the formation of an impermeable layer on the surface of the sulfur cathode, while the increase in DOL may alleviate this phenomenon. Similar work is described in [35, 36].

Dimethylene glycol dimethyl ether, with a glyme structure and low dielectric constant ($\varepsilon r = 7.9$) is another commonly used solvent for Li-S batteries. During the tests described in [31], a binary solvent consisting of TEGDME and DOL in a different volume ratio as well as a single TEGDME or DOL as a solvent for Li-S cells was used. The best results were obtained when the solvent with the composition TEGDME/DOL=25/75 (v/v) was used. The electrolyte solvation capacity was considered a key factor for achieving high electrochemical efficiency. These cells showed better results from TEGDME in combination with DOL in a 1:1 volume ratio as reported in [28].

The study of the ternary diglyme system (DGM)-DOL-DME shows that the capacity increases with a decrease in DOL content, and the average potential is higher with a lower DME content [36]. The lithium-sulfur cells provide an initial discharging capacity of 1303 $mAh\,g^{-1}$, using a TEGDME-based electrolyte at room temperature [22]. However, the capacity decreases significantly at a lower temperature due to the increase in viscosity of TEGDME when its temperature near the freezing point (−27°C). DOL has a lower freezing point at −45°C, so that its low-temperature performance is obviously better than that for TEGDME. The combination of TEGDME and DOL offers excellent capabilities [22, 37, 38], but a fully charged battery is difficult to achieve due to the high solid transfer rate of sulfides in low viscosity electrolytes [23, 39, 40]. To improve the ability of renewable Li-S batteries in charging, the phenomenon of polysulfide transfer can be inhibited by adding an adsorbent to a positive electrode or using a PEGDEG-based electrolyte that exhibits high viscosity, thus reducing the solubility of polysulfides in the electrolyte [23, 40]. Additives such as methyl acetate (MA), toluene, and gamma butyrolactone (GBL) were tested for their effect on the capacity and ability to work cyclically [22, 29]. Only toluene improves both the initial capacity and the ability to work cyclically due to its ability to create a stable contact between electrode/electrolyte [24, 29]. Imidazolium salts are introduced into mixed DME-DOL electrolytes, as reported, to improve the cyclic work capacity by increasing the electrochemical reaction of polysulfides and improving the negative stability of the lithium electrode [21]. Addition of tetrabutylammonium hexafluorophosphate (TBAPF6) to the electrolyte shows a comparable effect [41].

It seems that the suitable solvents for the Li-S cells' electrolytes are limited only to linear and cyclic ethers, such as dimethyl ether (DME) and 1,3-dioxolane (DOL).

In [11], we study the effect of solvents with 1 M lithium trifluoromethane sulfonate ($LiCF_3SO_3$) on the performance of Li-S batteries. As shown in Fig. 24A, two different voltage plateaus were observed at a level of about 2.4–2.1 V, for TEGDME and DOL/DME electrolytes, respectively. A small difference in the voltage window for both plateaus between the two systems was probably attributed to the difference in their bones [42]. Only one plateau was observed at 2.4 V for the electrolyte with the presence of propylene carbonates (PC)/ethylene carbonate (EC)/diethyl carbonate (DEC), which was because carbonates could react with the polysulfides by possible nucleophilic attack or ether or carbonyl atoms carbon on carbonate particles [43].

The first charging profiles for Li-S cells with different electrolytes are shown in Fig. 24B. The charging capacities were higher than 1000 mAh g^{-1} for TEGDME and DOL/DME. In contrast, there was hardly any capacitance for the PC/EC/DEC electrolyte. The cell with DOL/DME electrolyte solvents showed the best electrochemical efficiency, as shown in Fig. 25.

It is stated that linear DME provides higher polysulfide solubility and faster polysulfide kinetics while cyclic DOL forms a more stable solid electrolyte (SEI) interface on the surface of the Li anode, resulting in lower

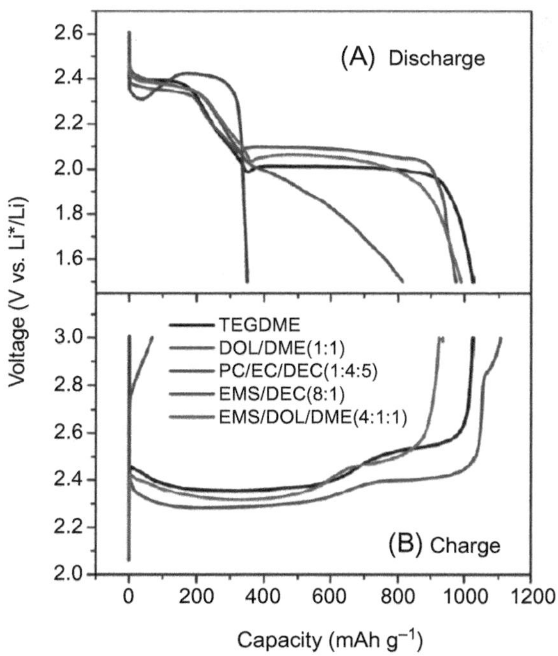

Fig. 24 Discharge profiles (A) and (B) charging in the first cycle for a Li-S cell in 1 M $LiCF_3SO_3$ with various solvents according to volume relations [11].

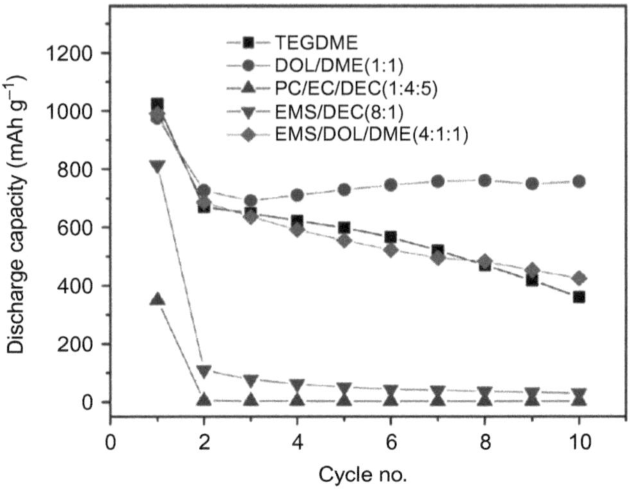

Fig. 25 Efficiency of cyclic work for Li-S cells with various solvents [11].

polysulfide dissolution and slower reaction kinetics. Although the dissolution of polysulfides causes parasitic redox-type transfer reactions, solvents with high solubility of polysulfides and low viscosity are much more advantageous for the cyclic performance of Li-S cells.

7.1.3 Lithium Salts

According to [44], the choice of lithium salt in electrolytes plays an important role. It is chosen based on chemical compatibility with polysulfides. Conventional lithium salts such as lithium hexafluorophosphate ($LiPF_6$) and lithium tetrafluoroborate ($LiBF_4$) are not suitable for the Li-S cell electrolyte due to the following reactions:

$$LiPF_6 + Li_2S_n \rightarrow LiPS_nF_4 + 2LiF \quad (15)$$

$$LiBF_4 + Li_2S_n \rightarrow LiBS_nF_2 + 2LiF \quad (16)$$

Lithium perchlorate ($LiClO_4$) is also not a good choice due to its oxidative properties, although no safety concerns have yet been reported. Until now, lithium bis (trifluoromethanesulfonate) sulfonamide (LiTFSI, $LiN(SO_2CF_3)_2$) is mainly used as the lithium salt of Li-S cells, because it provides higher ionic conductivity and less corrosion on the aluminum (Al) substrate [12, 45].

Lithium salts require high chemical and electrochemical stability, high solubility, and a high degree of dissociation in a solvent to ensure good ionic conductivity. Fig. 26 shows the typical lithium salts already used or

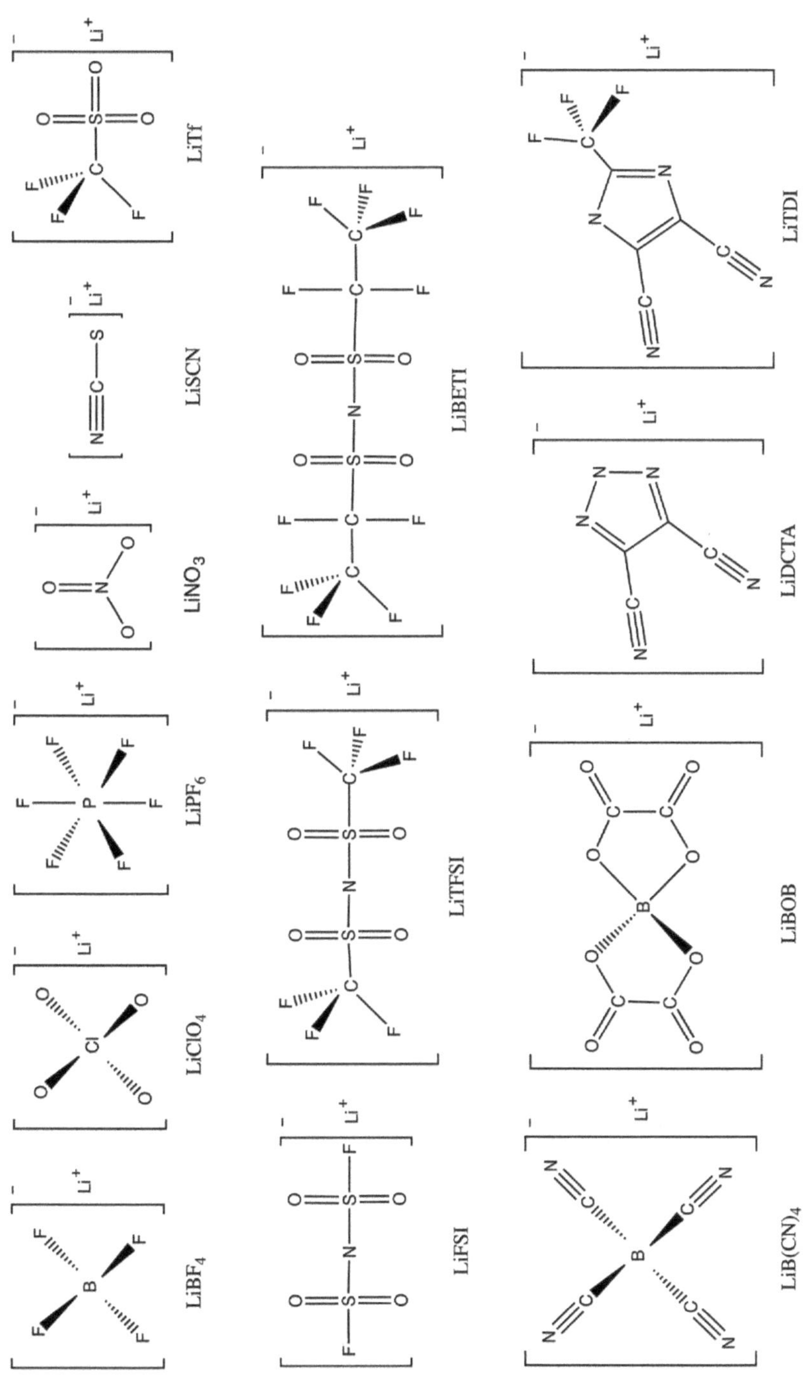

Fig. 26 Typical lithium salts already used or potentially applicable in Li-S cells [10].

potentially applicable in Li-S cells. Of these, the most commonly used lithium salts in the Li-S system are LiTFSI and LiTf [10, 42, 46], due to their high thermal stability, good compatibility with ether solvents and high dissociation ability. LiTFSI, and especially LiTf, are subject to Al-corrosion problems ranging from ~2.8 V [47] and are not recommended for use with most lithium-ion batteries. However, lower voltage of cel and milder conditions for the anode stability of Li-S batteries release LiTFSI and LiTf from this problem, as well as cause prolonged interest in some other salts with intermediate anode stability such as bis (fluorosulfonyl) imide (Li[N $(SO_2F)_2$], LiFSI) [48] and 4,5-dicyano-2-trifluoromethylimidazolide hydride (LiTDI) [49, 50]. One alternative is also lithium perchlorate ($LiClO_4$) [34]. The study of $LiClO_4$ is also described in [13]; its efficiency is usually lower than LiTFSI or LiTf, presumably because of the inability to produce a stable SEI on the lithium anode [51]. As for $LiPF_6$, a commonly used salt in lithium-ion batteries, it is rarely used for an ether system, due to the production of Lewis acids that break down the solvent [52]. To date, little research has been done into the effects of lithium salt on Li-S cell characteristics.

7.1.4 Electrolytes Based on Ionic Liquids

Recently, ionic liquid (IL) electrolytes have been used to improve the initial discharging capacity and cyclic capacity (due to reduced polysulfide dissolution) compared to electrolytes based on organic solvents such as PEGDME or DMC [53, 54]. The ability to work cyclically improves when the organic solvent is combined with liquid ionic electrolytes [55].

Ionic liquids were considered promising electrolytes for the Li-S battery due to thermal and chemical stability, with very good ionic conductivity [56]. These properties provide Li-S batteries with much greater safety in terms of avoiding leaks and gassing, a large window of electrochemical stability, and high solvent powers. Due to the lack of highly soluble ethers, PS dissolution and transfer effect can be effectively suppressed to maintain reduced losses of active material and high coulombic efficiency. However, the main disadvantages of ILs for the Li-S battery are higher viscosities than organic liquids and a higher cost compared to other conventional organic solvents. The first report on the use of pure IL as a matrix electrolyte for Li-S cells is contained in [57]. The new ILN-methyl-*N*-butyl-piperi-bis (trifluoromethanesulfonyl) imide (PP14-RTIL) working at room temperature, supported by 1 M LiTFSI, was used as an electrolyte for a Li-S battery. This cell showed an initial discharging capacity of 1.055 mAh g^{-1} with one displacement plateau at ~1.9 V, and more than 750 mAh g^{-1} after 10 cycles.

In [58], a Li-S system with a 0.64 M LiTFSI IL electrolyte based on tetravalent *N*, *N*-diethyl-*N*-methyl-*N*-(2-methoxyethyl) ammonium ammonia (DEME), and TFSI cations was compared with 0.98 M LiTFSI dissolved in the TEGDME electrolyte. Dissolution of PS was significantly inhibited due to the poor ability to form donors by anions $[TFSI]^-$. In this IL electrolyte, the electrochemical sulfur reduction took place mainly on the surface of the electrode through a solid phase process, which resulted in a better loss of the Li-S cells during cyclic operation. As reported in [59], extensive work was carried out, focusing on important roles of IL-anions in electrolytes. The results indicate that the nonionic sulfur had low solubility in all electrolytes evaluated, while the BF_4^- and anion $[FS]^-$ reacted irreversibly with PS on the surface of the electrode. The IL-based anion $[TFSI]^-$ and low viscosity polymers maintained significant chemical and electrochemical stability with PS during cyclic operation, allowing the best performance of a cell with sustained capacity >650 mAh g^{-1} and high coulombic efficiency >98% over 50 cycles. In [54] the behavior of a composite S-cathode/mesoporous carbon in Li-S cells immersed in an electrolyte based on 1 M 1-ethyl-3-methylimidazoliumbis (trifluoromethanesulfonyl) imide (EMIM TFSI)/LiTFSI was compared to the case of an electrolyte based on 1 M LiTFSI/PEGDME. The IL electrolytes exhibited both better cyclical capacity and discharging capacity because of the lower solubility of the PS found in the IL-based electrolyte.

Recently, as reported in [60], the so-called "quasi-IL" $[LiG_4]$ [NTf2] (G_4 = tetraethylene gly-col dimethyl ether, TEGDME), complex electrolytes of molten aglyme-Li salts, where the $[LiG_4]^+$ cation is an equimolar complex of G_4 and Li^+, was used. Corresponding cells with sulfur, supported on 3D ordered macroporous carbon, as a cathode material showed good reversibility, high capacity, and high coulombic efficiency.

In [61], an extended selection of [P1A3] [TFSI] (P1A3Dn-methyl-*N*-Allylpyrrolidinium) was reported for IL using an Si/C anode instead of metallic lithium. The Li-S cells produced a nominal voltage of 1.5 V and provided a high initial discharging capacity of 1.457 mAh g^{-1} at C/10 and maintained 670 mAh g^{-1} after 50 cycles. Although the ability to operate cyclic

Li-S cells has been improved by using the IL electrolyte, it also led to a low capacity to withstand nominal currents and low operating voltage [55]. To alleviate these problems, an electrolyte from the IL-organic mixture provided an alternative to classic IL. Early work on the IL-organic mixture was described in [21], introducing different amounts of EMIMBETI imidazolium salts and $BMIMPF_6$ into 0.5 M LiTFSI or 0.5 M $LiPF_6$ to DOL: DME (1:4 v/v). The best cell performance was obtained when 10% IL was used. Extensive work was carried out using many other ILs as additives in the same electrolyte based on DOL:DME (1:4 v/v) [62]. Li-S cells containing EMI and BMI showed a much higher capacity and cyclic capacity compared to Li-S LiTFSI–based cells, due to the lower resistance of charge transfer to the electrodes, while DMPI showed poor cell performance, probably due to its irreversible reaction with PS to form a nonconductive coating on the surface of the cathode. In conclusion, the electrolyte from the IL-organic mixture combines the advantages of the high conductivity of the inorganic solvent with the low solubility of PS in IL. The presence of an organic solvent was also planned to reduce the electrolyte viscosity, and give better wettability and higher conductivity for better electrochemical active sulfur kinetics. Based on these favorable properties, the IL-organic mixture is considered a promising electrolyte for Li-S batteries [63–65].

7.1.5 Carbonate-Based Electrolytes

The carbonate-based electrolyte, usually consisting of 1 M $LiPF_6$ dissolved in a binary solvent of ethylene carbonate (EC) and diethyl carbonate (DEC) in a 1:1 volume ratio, is most commonly used in lithium-ion batteries. However, it has been suggested that it is unsuitable for Li-S batteries due to a chemical reaction between nucleophilic sulfide anions and carbonates to form methylated thiolate and thioether during discharge [11]. Similar results have also been confirmed in [28, 66].

Carbonate solvents are generally considered to be inadequate electrolytes for Li-S batteries, but the latest investigations of PAN-S [67–69] or AC-S [70] composite electrodes have shown extremely high reversible capacities in mixtures: ethylene carbonate-dimethyl carbonate (WE-DMC) [67, 68, 71, 72] or propylene carbonate-ethylene carbonate-diethyl carbonate (PC-EC-DEC) [69, 70] containing 1MLiPF6. A sloping discharge profile was observed instead of a flat plateau in these systems [73]. It is uncertain whether carbonate solvents react with sulfur-containing discharges. Excellent ability to work cyclically and maximum efficiency may result from the reaction at the meeting of polysulfides or sulfides with carbonates, creating a protective layer on the surface of the cathode [73–75].

During the tests described in [75], 1 M $LiPF_6$ in PC/EC/DEC (1:4:5 v/v/v) was used as the electrolyte for the composite cathode sulfur/carbon spheres. The corresponding Li-S link showed only one lower plateau discharge voltage for about 1.8 V instead of two voltage plateaus at 2.3 and 2.1 V in a conventional ether-based electrolyte. This potential hysteresis was attributed to sulfur encapsulation in the narrow microporous carbon matrix. The composite containing 42% by weight of sulfur showed a long service life of 500 cycles at a high maintained capacity of over 80%, due to the limited electrochemical reaction inside the micropores.

In [76], compliance with the carbonate electrolyte was demonstrated when using metastable small S_4^{2-} sulfur molecules. S_4^{2-} was synthesized in a closed space of a conductive matrix of microporous carbon (MPC) to avoid generation of unfavorable long-chain PSs that are very reactive towards carbonates. These limited minor S_4^{2-} showed a single discharge plate at ~1.9 V, with a high specific capacity of 1670 mAh g^{-1} and an impressive cyclic cycling stability at 1149 mAh g^{-1} after 200 cycles. Based on this work, [77] investigated the mechanism of lithiation/deletion of the cathode from S2–4 comparing the electrochemical performance of the S_4^{2-} and S_4^{2-}/S_8 electrodes in the ether or carbon electrolyte. Fig. 27 shows the theoretically calculated dimensions of EC and DMC molecules, as well as sulfur lithiation processes in different pore structures. If the carbon micropores are small enough to prevent the permeation of solvent molecules, lithiation/deletion of S_4^{2-} sulfur occurs through a solid-solid interface process, which is inconsistent with a single plateau in the discharge potential profile. In this state, chemically reversible reactions between polysulfides are effectively avoided due to spatial obstructions, which ensures a high adaptability of the limited S_4^{2-} to the electrolytes. Corresponding links can show excellent ability to withstand maximum currents and cycle stability. Similar results were obtained using the S/porous carbon composite as a cathode in the carbonate electrolyte also obtained during the tests described in [78–80], obtaining good electrochemical results. However, one of the obvious problems associated with this strategy is the low sulfur content due to the small pore volume characterizing the micropores.

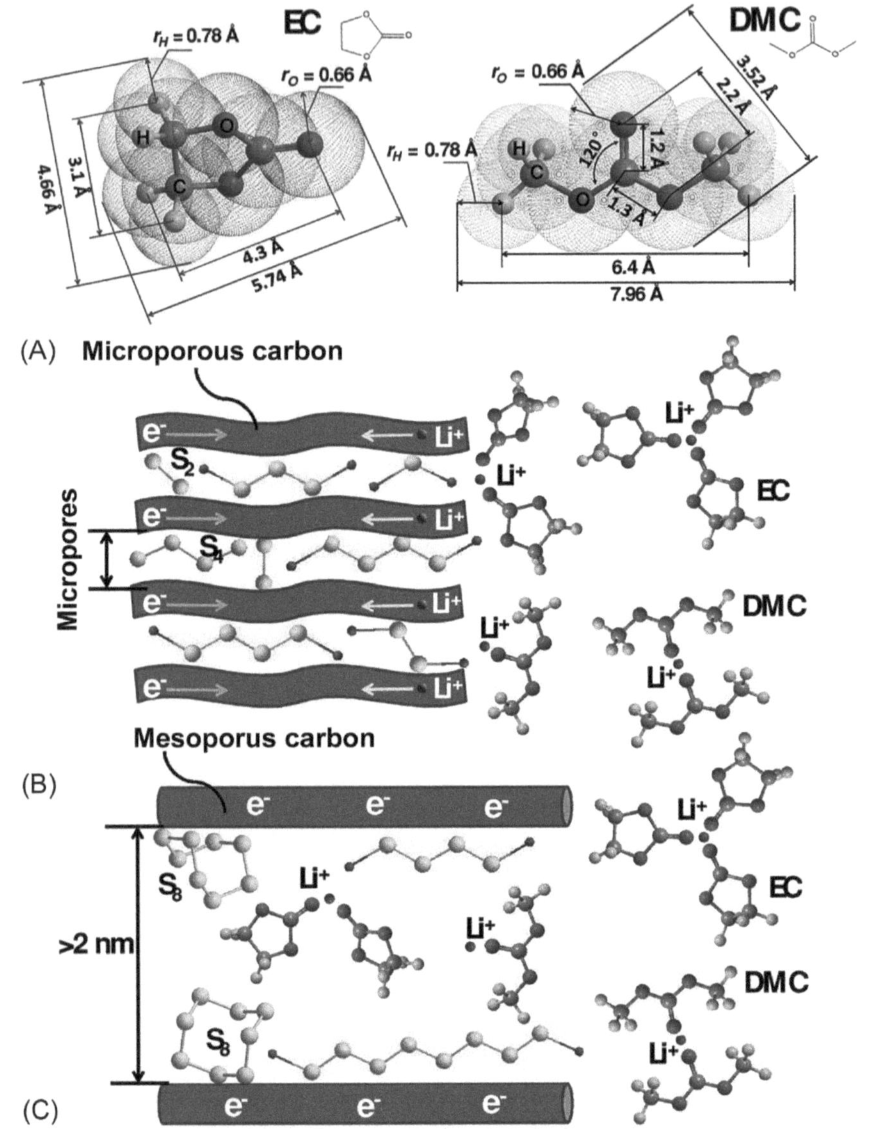

Fig. 27 Theoretically calculated dimensions of EC and DMC molecules (A), as well as sulfur lithiation processes in various pore structures (B, C) [77].

7.2 SOLID ELECTROLYTES

According to [81], the low solubility or insolubility of polysulfides led to the consideration of the use of solid electrolytes in Li-S batteries, with the promise of the elimination of polysulfide transfer, the mitigation of sulfur migration and the protection of the metallic lithium anode.

According to [10], solid electrolytes can effectively improve the performance of Li-S batteries.

Fig. 28 shows a typical Li-S discharge profile for all solid components. Compared to the typical two plateaus for a liquid organic electrolyte, Li-S batteries of all solid components show a single plateau at a relatively low potential of about 1.9V, which is caused by a single-phase reduction from S_8 to Li_2S in a solid-solid system, with deteriorated kinetics reaction and polarization.

According to [81], solid electrolytes, while conducting lithium ions, also function as separators in an electrochemical cell, electronically separating negative and positive electrodes. According to their nature, solid electrolytes used in Li-S cells are classified as polymer electrolytes, gel electrolytes, and glass/ceramic electrolytes.

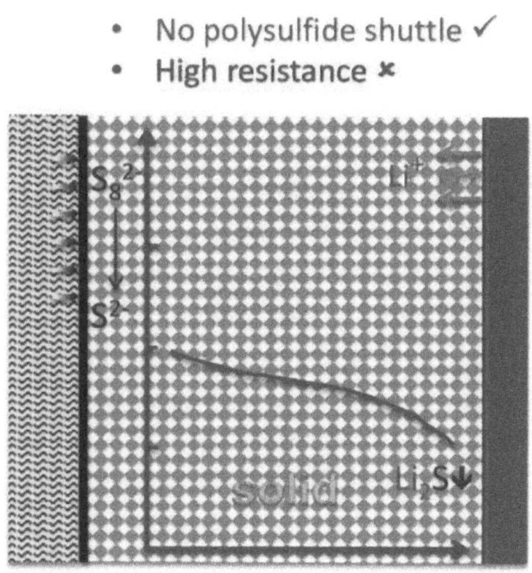

Fig. 28 A typical Li-S cell discharging profile with all solid components [10].

According to [82], compared to liquid electrolytes, solid-state electrolytes can both enhance safety and block the diffusion of polysulfides more effectively.

Various groups of solid electrolytes, including all-solid-state polymer electrolytes [71, 83], gel polymer electrolytes (GPEs; prepared by a hot-press [71], phase inversion [29, 84], or electrospinning [85] approaches), glass/ceramic electrolytes [86, 87], and fast ionic conductors [88–91] were utilized in Li-S batteries.

7.2.1 Solid Polymer Electrolytes

The solid polymer electrolyte is defined as a membrane having an ion transporting ability comparable to the dissociation of the lithium salt in typical liquid electrolytes [92].

The polymeric electrolyte, presented for the first time in [93], had favorable properties due to the lack of internal short circuits, electrolyte leakage, nonflammability, and suitability for applications in electrochemical devices [94].

A solid polymer electrolyte (SPE) usually consists of lithium salts dissolved in a high molecular weight polymeric host that function as solid solvents [95]. By applying an SPE to the Li-S cell system with all solid components, SPE can effectively suppress dissolution and diffusion of PS, as well as the transfer phenomenon, which greatly increases the ability to operate cyclic Li-S cells [96].

Many polymers, including poly (ethylene oxide) (PEO), PPO, PAN, PMMA, PVC, poly (vinylidene fluoride) (PVDF) and poly (vinylidene fluoride)-hexafluoropropylene (PVDF-HFP) have been accepted as solid hosts due to their high power of solvation and significant ion transport capacity. Among them, polymer host based on PEO is probably the best candidate, due to its great solvation power, complexing ability, and ion transport mechanism, directly related to alkaline lithium salts [97].

Various lithium salts (e.g., $LiClO_4$, LiTFSI, $LiBF_4$, $LiPF_6$) were used in combination with PEO, and their effect on ionic conductivity of the electrolyte is shown in Table 3 [98]. Also, according to [81], polymer electrolytes are lithium salts in PEO. The local coordination of lithium cations by ethylene oxide (EO) units in PEO confers a high ionic conductivity to the PEO polymer. When working at 80°C, Li-S batteries deliver more than 80% of the theoretical sulfur capacity with a lifetime of 400 cycles and an energy density of 200 Wh kg^{-1} [99]. Li-S batteries with polymer electrolytes

Table 3 Lithium-Ion Conductivity of the PEO-Li_xCPEs [98]

Electrolytes	Conductivity ($S\,cm^{-1}$)	Temperature (°C)
PEO-$LiBF_4$	1.12×10^{-4}	90
PEO-$LiPF_6$	2.69×10^{-4}	90
PEO-$LiCF_3SO_3$	8.38×10^{-5}	90
PEO-$LiCF_3SO_3$-10 wt% Ti_nO_{2n-1}	1.80×10^{-4}	90
PEO-$LiCF_3SO_3Li_2$ S-10 wt%ZrO_2	10^{-4}–10^{-3}	70
PEO-$Li(CF_3SO_2)_2$ N-10 wt%SiO_2	5×10^{-4}	70
PEO-$LiPF_6$-10 wt%TiO_2	2×10^{-5}	25
PEO-$LiClO_4$-10 wt% Al_2O_3	2.5×10^{-5}	25

can be optimized either for high discharging capacity or good cyclical capacity [26, 100], but it is difficult to optimize for both.

The PEO-$LiClO_4$ combination was tested in [101] as an electrolyte for the Li-S cell. The low ionic conductivity of the PEO-LIX electrolyte at room temperature (10^{-7} to $10^{-6}\,S\,cm^{-1}$) disqualified its use in Li-S cells. In relation to this problem, a recent promising strategy is the use of inorganic ceramic fillers, such as SiO_2, TiO_2, ZrO_2, Al_2O_3, and $LiAlO_2$ in the polymer matrix of the host, to increase the ionic conductivity of the SPE [102–110].

Also, in [111, 112] it was found that the ionic conductivity and stability at the interface of polymeric electrolytes is improved by adding ceramic fillers through high energy ball milling. The addition of Al_2O_3 improves the initial capacity, which was attributed to the stabilization of Li_2S by Al_2O_3. However, the loss of capacity is aggravated by the aluminum filler [83]. Similar results are obtained using TiO_2 and $LiAlO_2$ fillers [112, 113].

Recently, cyclicability of the Li-S battery with a polymer electrolyte has been improved by use of a $PEO_{20}LiCF_3SO_3$-ZrO_2 membrane swelled with 1M$LiPF_6$ and a saturated Li_2S solution in EC-DMC. Excellent properties have been demonstrated at 25°C and 60°C [71, 72, 114]. The components of this composite electrolyte play coordinated roles in achieving favorable properties: ZrO_2 improves mechanical properties and retention of liquid in the composite electrolyte, while Li_2S inhibits the dissolution of polysulfides. It is worth noting that high overpotential is observed in polymeric electrolytes. Consequently, Li-S batteries based on polymeric electrolytes have low energy efficiency. The reactivity of lithium with polymer, filler, or additives at elevated temperature and low ionic conductivity at room temperature call into question the use of polymeric electrolytes in Li-S batteries.

The nanofiller provides the advantage of reducing the crystallinity of the polymeric solvent and, as a result, produces an increase in ionic conductivity due to the amorphous phase. In addition, filler surface groups interact with oxygen atoms in PEO, and cross-linking polymer chains delay structural reorganization and increase the mobility of Li ions [115]. An important work is presented in [114], using ZrO_2 as an additive for electrolytes in Li-S batteries. The use of 10% by weight of ZrO_2 resulted in a significant improvement in ionic conductivity, transfer number of Li^+, and electrode/electrolyte contact stability, as well as the formation of a finely dispersed layer on the lithium surface [116, 117]. Corresponding Li-S cells maintained capacities up to $900\,mAh\,g^{-1}$ and sustained high coulombic efficiency up to 100% for 50 cycles. During the investigations described in [118], PEO 18LiTFSI-10 wt% SiO_2 was used as an electrolyte with a mesoporous carbon sphere with uniform channels (OMC) supporting sulfur as a cathode. The Li-S cells exhibited excellent cyclic performance with a reversible capacity up to $800\,mAh\,g^{-1}$ after 25 cycles. Although SPE has been successfully optimized by inorganic additives, nominal performance remains low (usually no more than C/20); moreover, relatively high temperature is necessary for adequate ionic conductivity.

7.2.2 Gel Polymer Electrolytes

Gel polymer electrolytes (GPEs), in which the liquid component is tightly trapped in the polymer matrix, were developed as a compromise approach to the low ionic conductivity of the SPE [119]. The volume of the membrane usually consists of a combined micropore to ensure an adequate amount of liquid electrolyte absorption. The ionic conductivity is to a large extent dependent on the properties of the liquid electrolyte, or otherwise the Li^+ transfer takes place mainly in the polymer membrane, with insufficient absorption of the liquid [120].

According to [81], gel electrolytes have better ionic conductivity than polymeric electrolytes. According to [82], they also possess higher ionic conductivities, better interfacial property, and more controllable functional groups compared with all solid-state electrolytes. In gel polymer electrolytes, the functional groups on the polymer backbones play an essential role in the interaction with polysulfides and the electrochemical performance of batteries.

Many types of polymer films, such as those based on PEO, PVDF, and PVDF-HFP, have been developed and characterized for Li-S cells

[68, 69, 71, 121–123]. Wang et al. investigated a gel electrolyte poly (vinylidene fluoride-hexafluoropropylene) (PVDF-HFP) with plasticizers, such as EC-DMC, and PC-EC-DEC for Li-S batteries, at room temperature. The ionic conductivity has significantly improved to about $1.2 \times 10^{-3}\,S\,cm^{-1}$, which is much higher than that for a PEO-based polymeric electrolyte. High discharging capacity and use of sulfur are obtained with PAN-S [69, 78] and C-S [67, 70] composite electrodes. However, much less use of sulfur is obtained when PVDF [122] or PVDF-HFP [18] are combined with TEGDME as a plasticizer. When TEGDME combines with KE as a plasticizer for PVDF-HFP microporous gel electrolytes, better results are observed in lithium-sulfur batteries [124].

During the tests described in [122], a polymer matrix based on PVDF was used in combination with LiTf/tetraglyme as an electrolyte for Li-S cells. Two plates were observed on the first discharge with high capacity up to $1268\,mAh\,g^{-1}$. However, the upper plateau disappeared at the second discharge with the disappearing capacity of $1.028\,mAh\,g^{-1}$, indicating the hindering effect of GPE in dissolving long-chain PS. In [71], the construction of a metallic system without mercury using the anode Sn/C and the cathode Li_2S/C was noted. A gel-type polymer membrane formed by the capture of $LiPF_6$/ECCDMC in a PEO/$LiCF_3SO_3$ polymer matrix was used as an electrolyte with a dispersed ceramic zirconium filler as an addition to improve mechanical properties. The impedance tests indicated low resistance and high conductivity close to $10^{-2}\,S\,cm^{-1}$. The polymer cell obtained showed a high initial capacity of over $1200\,mAh\,g^{-1}$ and good storage capacity with a C/20 load during cyclic operation. GPE functioned as a barrier, preventing dissolution of the PS, and the Sn/C anode was intended to eliminate the corrosion risk of the anode for the liquid electrolyte. Due to the combination of SPE and liquid electrolyte, higher ionic conductivity and better reaction kinetics were obtained by using GPE at the expense of the reemergence of PS dissolution and the transfer effect. However, more in-depth and systematic research must be carried out, such as on how GPE and GPE composition influence PS solvation characteristics as well as changes in the pore structure and crystallinity of the polymer to improve Li^+ transport.

Seh et al. [125] found that sulfur species exhibited the strongest interaction with oxygen donor atoms in ester (R–COO–R′) groups. Park et al. [126] experimentally verified that ester groups with a high spatial density can strongly bind polysulfides. The electron-donating ability and spatial density of the functional groups determined the strength of Li-polymer interactions and the corresponding electrochemical performances.

Liu et al. [127] synthesized a pentaerythritol tetraacrylate (PETEA)-based gel polymer electrolyte with an ultrahigh ionic conductivity ($1.13 \times 10^{-2}\,S\,cm^{-1}$) and beneficial functional groups (R–COO–R′) for Li-S batteries, which achieved high compatibility with solid sulfur cathodes and successfully inhibited the diffusion of polysulfides, attributed to the novel electrolyte and rational cell configuration design. The electrochemical performance of Li-S batteries strongly depended on the ionic conductivity of the gel polymer electrolyte and its special functional groups. Accordingly, the PETEA monomer had a symmetrical star structure with four C=C bonds in each molecule, which provided the polymerized PETEA with a higher crosslink density compared with other monomers. Such rigid polymerized PETEA matrix negligibly inhibited the ion mobility in gel polymer electrolytes and allowed high ionic conductivity and high rate performance in Li-S cells. This in situ synthetic method avoided the major sacrifice of specific energy in Li-S cells. Afterwards, the acrylate-based gel polymer electrolyte was further in situ gelated into a polymethyl methacrylate (PMMA)-based electrospun nanofiber network. The structural similarity and synergetic compatibility between the nanofiber network and gel polymer electrolyte provided a strong uptake ability and enhanced ionic conductivity of the electrolyte, combined with an ester-rich structure, which led to improved electrochemical performances [128].

7.2.3 Nonpolymer Electrolytes

Due to the low performance of cells using solid electrolytes based on polymers, nonpolymer electrolytes offer a promising strategy for all—solid component Li-S cells, considering their advantages related to high thermal and chemical stability of the Li anode in the ambient atmosphere [129–132]. A large number of inorganic semiconductor electrolytes were tested for Li-S cells as well as lithium-ion batteries, including phosphates of the LiSICON type, perovskite type $La_{(2/3)x}Li_{3x}TiO_3$ (LLT), Li_3N and Li_4SiO_4 [86, 87, 91, 133, 134]. The list of nonpolymeric electrolytes in comparison to other electrolytes, including polymeric electrolytes, ionic liquids, and gel electrolytes, is presented in Fig. 29.

The family of sulfide compounds that form a solid amorphous electrolyte with high conductivity Li^+ is one of the most important inorganic electrolytes in the solid state. Several significant works were carried out on the glass electrolyte Li_2S-P_2S_5 [136–140]. Agostini et al. [141] found that a solid-state Li-S cell with the Li_2S–P_2S_5 electrolyte could achieve the merits

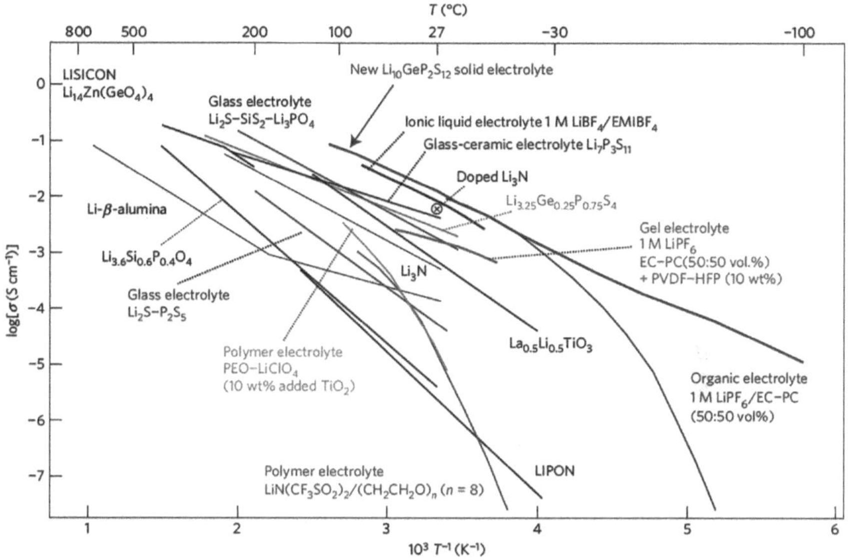

Fig. 29 Thermal evolution of the ionic conductivity of the $Li_{10}GeP_2S_{12}$ phase at once for solid lithium electrolytes, polymer electrolytes, ionic liquids and gel electrolytes [135].

of high-specific energy, safety, and low cost. Afterwards, Yamada et al. [142] demonstrated that in the all-solid-state Li-S battery using a glass-type P_2S_5-Li_2S electrolyte, the solid electrolyte component physically prevents polysulfide migration.

Mechanically ground glass Li_2S-$20P_2S_5$ (mol%) was shown to have an extremely high conductivity up to 10^{-3} S cm^{-1} at room temperature, due to the formation of Li_7PS_6 [143], described as "superionic" and crystallographic analogs of thio-LISICON [144]. The Li_2S-P_2S_5 ceramic electrolyte was successfully used in the Li-S system. Corresponding cells with the S/CuS cathode retained the capacity of over 650 mAh g^{-1} for 20 cycles [87]. Further development was obtained using a sulfur-carbon composite electrode and Li_2S-P_2S_5 ceramic glass electrolyte. Excellent cell performance was maintained at a reversible capacity of over 850 mAh g^{-1} for 200 cycles of charging/discharging below 1.3 mA cm^{-2} at room temperature, and even close to 500 mAh g^{-1} at 20 mA g^{-1} when the temperature rises to 80°C [86]. Recently, the related compound $Li_{10}GeP_2S_{12}$ showed the highest ionic conductivity of about 10^{-2} S cm^{-1} at room temperature with good potential for use in the Li-S system, but the relatively limited reversibility of metallic lithium remains to be overcome [135]. Also, cells using the Li-In alloy as a negative electrode and $Li_{10}GeP_2S_{12}$ as a fixed electrolyte were assembled

[145]. The cathode material was prepared by mechanical-chemical treatment of a mixture of $Li_{1.5}PS_{3.3}$, sulfur and conductive carbon. The obtained cathode showed a reversible capacity of 1600 mAh g^{-1} (sulfur-based), after 100 cycles at a load of 1C at 25°C, showing a high power density above 11,000 W kg^{-1} at 50% SOC. Recently, the Liang group at the Oak Ridge National Laboratory reported nanoporous β-Li_3PS_4 with high conductivity up to 1.6×10^{-4} S cm^{-1} at room temperature as a result of a liquid state reaction [146]. Reversible lithium deposition and perfect settlement against lithium dendrites were identified by CVs and cells operating cyclically at 25°C. In the subsequent development of this material, nanometric Li_2S molecules coated with β-Li_3PS_4 were obtained as the active substance for the cathode [147]. For the phosphorothioate also used as the electrolyte, metallic lithium, as the anodes, the assembled cells provided a significant 1.216 mAh g^{-1} (sulfur-based) capacity on the first discharge, but at the same time decreased by about 30% during 100 cycles at 60°C. It should be noted that the small coal surface used may result in a lack of extensive interaction between sulfur/sulfide and functional support. Another use of this material is in the form of P_2S_5 additive to a conventional liquid electrolyte, by forming β-Li_3PS_4 on the electrolyte surfaces, ensuring good ion transport, protection of the anode from PS, and increased solubility for sulfides. In [148], the use of a new class of sulfur-rich polysulfide lithium phosphate as the cathode for the Li-S battery has been reported. These compounds showed high conductivity of Li 10^{-4} to 10^{-5} S cm^{-1} ions at room temperature. Excellent properties were demonstrated when the cathode material was combined with a solid Li_3PS_4 electrolyte to obtain a reversible capacity of over 1200 mAh g^{-1} at C/10 for 300 cycles at 60°C.

Li_2S-SiS_2 and Li_2S-P_2S_5 ceramic glasses exhibit high ionic conductivity S 10^{-4} cm^{-1} at room temperature and are promising electrolytes in all Li-S batteries with all-solid components. Conductivity is improved by crystallization of $80Li_2S \bullet 20P_2S_5$ (mol%) by mechanical grinding and reaches 10^{-3} S cm^{-1} [87, 138, 149]. The high capacity of 650 mAh g^{-1} based on CuS mass is maintained after 20 cycles, when CuS is used as a positive electrode and metallic In (indium) as a current collector for lithium deposition [137]. The initial capacity of 490 mAh g^{-1} is achieved based on the weight of Li_2S-Cu. The capacity of 350 mAh g^{-1} is maintained after 20 cycles [138]. Glass $60Li_2S \bullet 40SiS_2$ has a lithium-ion conductivity of 1.8×10^{-4} S cm^{-1} at room temperature [150, 151]. The high capacity of 1080 mAh g^{-1} is obtained using a $60Li_2S \bullet 40SiS_2$ glass electrolyte [137]. Some new lithium-ion conductors have been developed recently, such as thio-LISICON ($Li_{3.25}Ge_{0.25}P_{0.75}S_4$) and $3Li_4SiS_4 \bullet 0.5La_2S_3$ [152, 153]. Reversible 900 mAh g^{-1} capacity is obtained when thio-LISICON is used as a solid electrolyte [134]. Excellent cyclical performance has been achieved at extremely low currents. Li-S batteries with all solid components have 1.4–1.5 V discharge potential, which is significantly lower than Li-S batteries based on liquid or polymer electrolytes. High resistance at the interface between Li-S batteries and all solid components requires further improvement to enable the practical use of these batteries.

Han et al. [90] reported a bottom-up method to synthesize nanocomposites by dissolving Li_2S as the active material, polyvinylpyrrolidone (PVP) as the carbon precursor, and Li_6PS_5Cl as the solid electrolyte in ethanol, followed by coprecipitation and high-temperature carbonization processes. The attained nanocomposites could be used as an active material and electrolyte simultaneously. When the $80Li_2S{\bullet}20P_2S_5$ glass ceramic was further used as the solid electrolyte in the composite electrode, acceptable electrochemical performances were attained.

Wang et al. [89] presented a comprehensive assessment of the durability of the all-solid-state electrolyte $Li_{1+x}Ti_{2-x}Al_x(PO_4)_3$ in contact with a polysulfide solution and in Li-S cells. It is undesirable that when $Li_{1+x}Ti_{2-x}Al_x(PO_4)_3$ is in contact with the polysulfide solution, Li_2CO_3 is formed on the particle surface, as it increases the interfacial resistance. After lithium ion insertion into the framework, the crystal expands in an anisotropic way, which weakens the crystal bonds, causes fissures and resultant cracks in the ceramic, corrodes the grain boundaries by the polysulfide solution, and leaves unfavorable pores. Although they possess safety merits, solid electrolytes still suffer from limited ionic conductivities ($<3 \times 10^{-3}\,S\,cm^{-1}$), low stability, and large electrode/solid electrolyte interfacial resistances, resulting in a rapid capacity decay and large polarization for solid Li-S batteries.

7.3 ELECTROLYTE ADDITIVES

Since the solvents and lithium salts of Li-S batteries are solid, the focus is only on electrolyte additives that can affect the volume properties of the electrolyte, such as ionic conductivity, viscosity, wettability, etc. [25, 29, 36, 38, 154–156]. It has been shown that various additives protect the Li electrode, creating a protective surface layer and suppressing the transfer effect. Most Li-S battery electrolytes contain small amounts of additives to protect the lithium anode and increase solubility and PS stability. Several additions are discussed in the following sections.

7.3.1 Lithium Nitrates

$LiNO_3$ has become the most-used addition in the Li-S battery since the first application in 2008. $LiNO_3$ functions in the Li-S system are described in detail in [157]. The results showed that the use of $LiNO_3$ leads to a lower impedance of the Li film, and the products of its reduction Li_xNO_y contribute to the formation of a suitable surface coating that protects Li from further reaction with PS. However, the $LiNO_3$ study, described in [158], indicated that the beneficial passivation of the anode is accompanied by a less desirable effect in the cathode. During the first discharge, it was found that $LiNO_3$ was irreversibly reduced on the surface of the carbon, and the reduction products negatively affected the reversibility of the active sulfur redox, with a negative effect on the capacity and cyclic performance of the Li-S battery. It was recommended not to discharge the cell below 1.6 V to avoid the problem of $LiNO_3$ reduction [32].

One of the most commonly used additives in Li-S electrolytes is lithium nitrate ($LiNO_3$), which can create a stable SEI passivation layer on the surface of Li electrodes [159]. The protective film effectively improves the stability of the Li anode and suppresses the effect of polysulfide transfer and self-discharge of Li-S cells, increasing the capacity as well as coulombic efficiency of the cells [160].

Aurbach et al. reported that the additional $LiNO_3$ in the electrolyte may improve the discharging capacity, which increased from ~650 to ~1150 mAh g^{-1}, as shown in Fig. 30 [157]. It was obvious that the upper

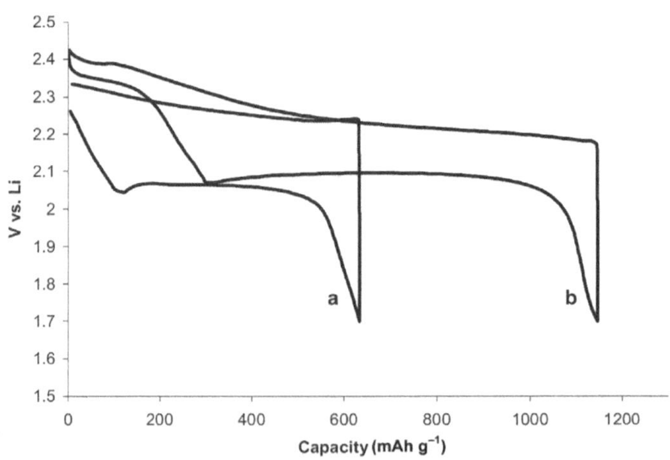

Fig. 30 Typical Li-S cells discharging/charging profiles (A) without and (B) with the addition of $LiNO_3$ [157].

plateau of discharge was significantly shortened in the absence of $LiNO_3$, which was attributed to the loss of long chain polysulfides, reducing the use of active materials.

The mechanism of action of the SEI induced by $LiNO_3$ to prevent the transfer effect in Li-S batteries was subject to further research, described in [159]. Based on the characteristics obtained from the scanning electron microscope (SEM) and photoelectron spectroscopy (XPS), it is believed that both $LiNO_3$ and polysulfides play an important role in the formation of stable SEI layers on the Li electrode, shown in Fig. 31. The SEI film is produced and increases during cyclic work in the electrolyte with $LiNO_3$ as an additive that is used due to the strongly oxidizing nature of $LiNO_3$.

7.3.2 Polysulfides

Polysulfides were also used as additives to the Li-S electrolyte due to their participation in the formation of SEI anodes [161, 162]. The concept of a Li-S liquid battery or cathode represents the extreme approach of this strategy, while most of the related works still focus on PS as an addition to Li-S cells, together with mesoporous S/C cathodes [161, 163–165]. In [162], research is presented on the formation of a durable solid two-layer contact on the metallic lithium surface during cyclic operation of symmetric cells Li/Li and Li/S with 0.2 M Li_2S_6 as a cosalt in electrolyte 0.8 M LiTFSI DOL: DME (1:1 v/v). The top layer consisted of LiTFSI decomposition products due to the initial contact with the lithium anode. The inner layer consisted of reduction products because of a chemical reaction between metallic lithium and PS in the electrolyte. In [162], it was suggested that the inner layer effectively prevents continuous LiTFSI decomposition, and the upper layer cannot effectively inhibit the PS transfer.

Long-chain polysulfides are used as an additive in the electrolyte to eliminate the formation of insoluble Li_2S_2 and Li_2S within the carbon matrix and the formation of a protective layer SEI on the electrode Li. As shown in Fig. 31, for an electrolyte with the addition of polysulfides, the stable SEI film is also produced during cyclic operation, which consists mainly of solid Li_2S_2 and Li_2S.

Two additions were also used together, $LiNO_3$ and lithium polysulfides, to investigate the formation of the SEI film in Li-S cells [159]. The results obtained indicate that the SEI film has two sublayers. First, the reduced $LiNO_3$ and polysulfide products are deposited on the Li electrode, resulting in a dense bottom layer consisting of Li_2S_2, Li_2S, and LiN_xO_y. The next layer

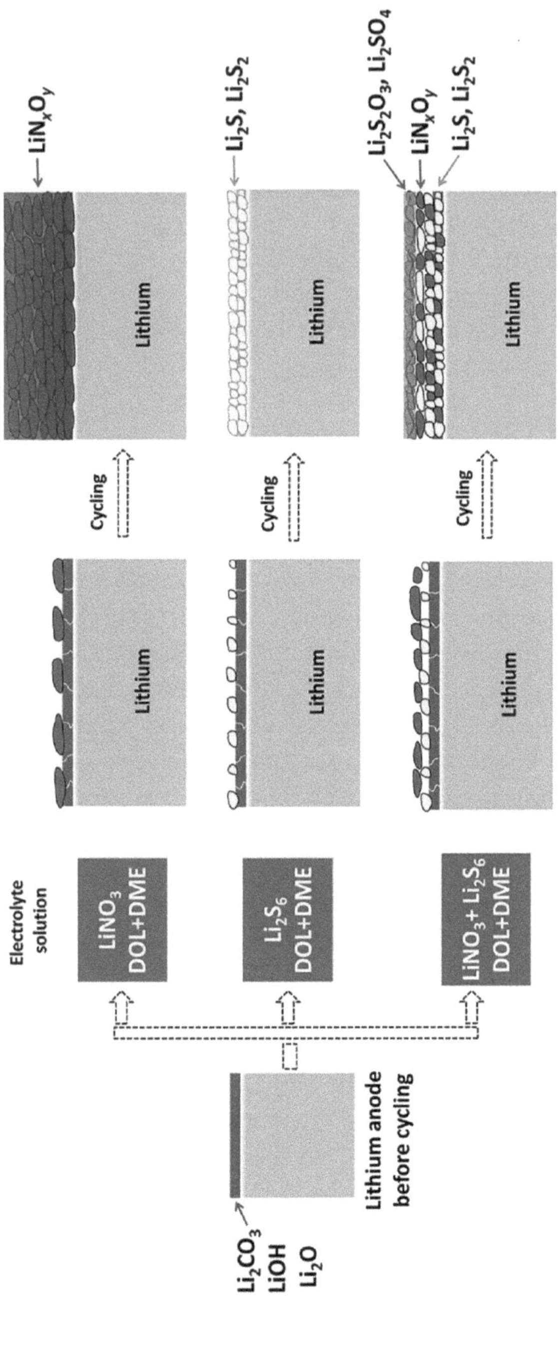

Fig. 31 Illustration of the behavior of a surface film on a lithium anode working periodically with various additives [159].

is derived from the reaction between $LiNO_3$ and lithium polysulfides, forming stable $Li_2S_2O_3$ and $LiSO_4$ compounds. This upper layer can block the direct contact between the polysulfides in the electrolyte solution and the reducing elements at the Li electrode during cyclic operation.

7.3.3 Phosphorus Pentasulfide

Another additive, phosphorus pentasulfide (P_2S_5), is described in [166]. The P_2S_5 function was introduced for two reasons:

(1) P_2S_5 promotes the dissolution of Li_2S and reduces the loss of capacity caused by Li_2S precipitation,

(2) P_2S_5 is involved in the formation of the passivation layer on the lithium surface and therefore eliminates the phenomenon of polysulfide transfer.

The Li-S test cell showed a high reversible capacity of 900–1350 $mAh\,g^{-1}$ and high coulombic efficiency $\geq$90% for 40 stable cycles at C/10.

It was found that P_2S_5 can be used as a new additive to the organic electrolyte for Li-S batteries, for which it has been shown that P_2S_5 can react with the Li electrode to create a thick passivation layer between the Li electrode and the electrolyte [166]. This passivation layer can help prevent polysulfide access and reaction with the Li electrode.

As shown in Fig. 32, a capacity of ~900 $mAh\,g^{-1}$ can be achieved after 40 cycles with the addition of P_2S_5. In contrast, the cell capacity without the addition of P_2S_5 was less than 300 $mAh\,g^{-1}$. This poor performance of cyclic

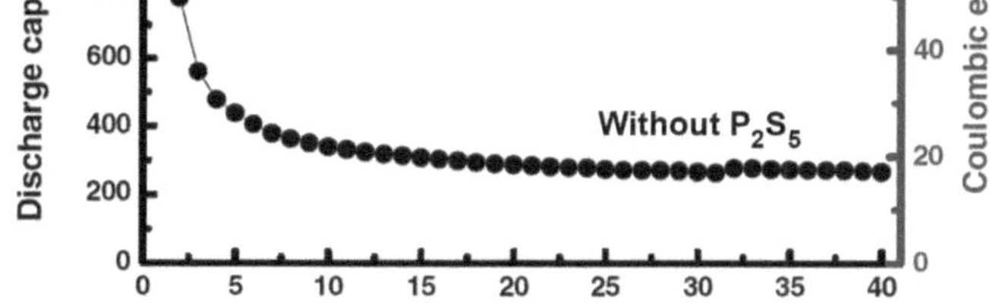

Fig. 32 Capacity of cyclic operation of Li-S batteries with and without P_2S_5 [166].

work was probably due to the direct reduction of polysulfides on the surface of the Li electrode and the formation of an SEI film consisting mainly of Li_2S, which is a poor conductor of lithium ions.

7.4 ELECTROLYTE DECOMPOSITION MECHANISMS

The important, although undesirable in the case of Li-S cells, phenomenon of electrolyte decomposition was studied, among other things, in [167]. That decomposition was monitored by fitting a pressure sensor and applying mass spectroscopy (MS) to a Li-S cell. The pressure increased during discharge due to the release of gases such as carbon disulfide (CS_2) and nitrogen, produced by the decomposition of the electrolyte. N_2 gas was produced due to the decomposition of lithium nitrate ($LiNO_3$) in the electrolyte, an additive which was widely used to suppress the PS redox shuttle. In contrast, electrolytes without it did not show any N_2 gas evolution and, therefore, $LiNO_3$ was not beneficial since it promoted decomposition of the electrolyte.

In [168], the stability of ether-based electrolytes (1 M LiTFSI, 0.25 M $LiNO_3$ in DOL:DME) for Li-S batteries was studied, concerning the effect of dissolved oxygen. The performance of two different electrolyte solvents, i.e., 1,2-dimethoxyethane and its mixture with 1,3-dioxolane (DME:DOL, 1:1 v/v), was characterized in cells assembled in a dry air environment.

Linear solvents, such as DME, are more resistant to decomposition by oxygen species, while DOL undergoes catalytic ring-opening reactions, which happens when O_2 is present or not, leading to fast solvent depletion. The cyclability of sulfur is affected by this decomposition of DOL (which starts at the anode side, before the cell is cycled). One proposed path of decomposition is radical-catalyzed isomerization of DOL, leading to the formation of esters, which react with the polysulfides. Such reactions increase the formation of trisulfur radical anion, electrochemically detected as the increased capacity of the sloping region during discharge. While this reaction eventually consumes all available active material, it also increases the capacity during the first discharge compared to ordinary electrolytes. For the cycled cathodes the thyocarboxylate species occurred.

DOL was shown to aid in the SEI formation in cells containing $LiNO_3$ as additive. DOL polymerizes on the anode, forming a more stable SEI than that formed by only nitrate, as determined by EIS. Still, cells with and without DOL in the electrolyte provided similar specific capacity and capacity retention upon cycling. At least for the limited number of cycles, the presence of DOL is not necessary to obtain good capacity retention. Those findings were also relevant to the Li-O_2 batteries, where these ether-based electrolytes are also used.

During studies described in [169], the degradation mechanism due to the decomposition of organic electrolytes during charge and discharge cycling of Li/MoS_2 cells was investigated. It was found that the observed gases were produced through two types of electrolyte decomposition: the electrochemical reaction, which was dependent on the charge and discharge voltage, and the chemical reaction, which was independent of the discharge voltage. This decomposition of electrolytes occurred under a discharge voltage lower than 1.4 V. The operating conditions with a lower discharge rate and a deeper depth of discharge (DOD) accelerated the electrochemical decomposition. The rate of gas production by chemical decomposition, which is independent of the discharge condition, was estimated to be about 0.1 mL/h.

During studies described in [170], a simple mixture of MoS nanoparticles, conductive carbon, and polymeric binder was used for the elementary cathode. After full lithiation of the MoS_2, Li_2S nanoparticles were produced into a highly stable polymeric gel-like film (derived from the electrolyte decomposition), which tightly confines sulfur-based species into it.

A cell containing a lithiated silicon/carbon composite anode assembled vs. the MoS_2 cathode was also investigated. The electrolyte consisted of 1 M $LiPF_6$ in ethylene carbonate (EC)/dimethyl carbonate (DMC) (1:1 v/v) (LP30); 10% v/v of fluoroethylene carbonate (FEC) was added to the LP30 electrolyte to stabilize the Si anode.

It was stated, that during discharging of the cell the solid electrolyte interface (SEI) film is usually formed at potentials below 0.8 V vs. Li/Li^+ because of electrolyte decomposition. At the discharge potential of 0.01 V vs. Li/Li^+, various undesired side phenomena can occur—large volume expansion of the particles upon lithiation with partial damage of the formed SEI and the rapid depletion of electrolyte compounds with a continuous growth of the SEI film—which isolates the active material from the conductive current collector.

REFERENCES

[1] S.S. Zhang, Improved cyclability of liquid electrolyte lithium/sulfur batteries by optimizing electrolyte/sulfur ratio, Energies 5 (12) (2012) 5190–5197.

[2] L. Carbone, M. Gobet, J. Peng, M. Devany, B. Scrosati, S. Greenbaum, J. Hassoun, Comparative study of ether-based electrolytes for application in lithium–sulfur battery, ACS Appl. Mater. Interfaces 7 (25) (2015) 13859–13865.

[3] R.D. Rauh, K.M. Abraham, G.F. Pearson, J.K. Surprenant, S.B. Brummer, A lithium/dissolved sulfur battery with an organic electrolyte, J. Electrochem. Soc. 126 (4) (1979) 523–527.
[4] M. Morita, S. Aoki, Y. Matsuda, ac imepedance behaviour of lithium electrode in organic electrolyte solutions containing additives, Electrochim. Acta 37 (1992) 119–123.
[5] H. Yamin, E. Peled, Electrochemistry of a nonaqueous lithium/sulfur cell, J. Power Sources 9 (1983) 281–287.
[6] R.D. Rauh, F.S. Shuker, J.M. Marston, S.B. Brummer, Formation of lithium polysulfides in aprotic media, J. Inorg. Nucl. Chem. 39 (1977) 1761–1766.
[7] Y. Jung, S. Kim, B.-S. Kim, D.-H. Han, S.-M. Park, J. Kwak, Effect of organic solvents and electrode materials on electrochemical reduction of sulfur, Int. J. Electrochem. Sci. 3 (2008) 566–577.
[8] V. Kolosnitsyn, E. Kuzmina, E. Karaseva, Reasons for the effect of the amount of electrolyte on the performance of lithium–sulfur cells, ECS Trans. 19 (2009) 25–30.
[9] G. Li, Z. Li, B. Zhang, Z. Lin, Developments of electrolyte systems for lithium–sulphur batteries: a review, Front, Energy Res. 3 (5) (2015), https://doi.org/10.3389/fenrg.2015.00005.
[10] J. Scheers, S. Fantini, P. Johansson, A review of electrolytes for lithium-sulphur batteries, J, Power Sources 255 (2014) 204–218, https://doi.org/10.1016/j.jpowsour.2014.01.023.
[11] J. Gao, M.A. Lowe, Y. Kiya, H.D. Abruña, Effects of liquid electrolytes on the charge-discharge performance of rechargeable lithium/sulfur batteries: electrochemical and in-situ X-ray absorption spectroscopic studies, J. Phys. Chem. C 115 (2011) 25132–25137, https://doi.org/10.1021/jp207714c.
[12] S.S. Zhang, Liquid electrolyte lithium/sulfur battery: fundamental chemistry, problems, and solutions, J. Power Sources 231 (2013) 153–162, https://doi.org/10.1016/j.jpowsour.2012.12.102.
[13] H.-S. Kim, C.-S. Jeong, Electrochemical properties of binary electrolytes for lithium-sulfur batteries, Bull. Kor. Chem. Soc. 32 (2011) 3682–3686, https://doi.org/10.5012/bkcs.2011.32.10.3682.
[14] G.Y. Xu, B. Ding, J. Pan, P. Nie, L.F. Shen, X.G. Zhang, High performance lithium-sulfur batteries: advances and challenges, J. Mater. Chem. A 2 (2014) 12662–12676, https://doi.org/10.1039/C4ta02097a.
[15] D. Aurbach, O. Youngman, P. Dan, The electrochemical-behavior of 1,3-dioxolane-LiCLO4 solutions. 2. Contaminated solutions, Electrochim. Acta 35 (1990) 639–655, https://doi.org/10.1016/0013-4686(90)87056-8.
[16] D. Aurbach, O. Youngman, Y. Gofer, A. Meitav, The electrochemical-behavior of 1,3-dioxolane-LiCLO4 solutions. 1. Uncontaminated solutions, Electrochim. Acta 35 (1990) 625–638, https://doi.org/10.1016/0013-4686(90)87055-7.
[17] Y. Gofer, Y.E. Ely, D. Aurbach, Surface-chemistry of lithium in 1,3-dioxolane, Electrochim. Acta 37 (1992) 1897–1899, https://doi.org/10.1016/0013-4686(92)85096-4.
[18] H.S. Ryu, H.J. Ahn, K.W. Kim, J.H. Ahn, J.Y. Lee, E.J. Cairns, Self-discharge of lithium–sulfur cells using stainless-steel current-collectors, J. Power Sources 140 (2005) 365–369.
[19] S.-I. Tobishima, H. Yamamoto, M. Matsuda, Study on the reduction species of sulfur by alkali metals in nonaqueous solvents, Electrochim. Acta 42 (1997) 1019–1029.
[20] H. Yamin, A. Gorenshtein, J. Penciner, Y. Sternberg, E. Peled, Lithium sulfur battery oxidation/reduction mechanisms of polysulfides in THF solutions, J. Electrochem. Soc. 135 (1988) 1045–1048.

[21] S. Kim, Y.J. Jung, S.J. Park, Effects of imidazolium salts on discharge performance of rechargeable lithium–sulfur cells containing organic solvent electrolytes, J. Power Sources 152 (2005) 272–277.
[22] H.-S. Ryu, H.-J. Ahn, K.-W. Kim, J.-H. Ahn, K.-K. Cho, T.-H. Nam, J.-U. Kim, G.-B. Cho, Discharge behavior of lithium/sulfur cell with TEGDME based electrolyte at low temperature, J. Power Sources 163 (2006) 201–206.
[23] H.S. Ryu, H.J. Ahn, K.W. Kim, J.H. Ahn, K.K. Cho, T.H. Nam, Self-discharge characteristics of lithium/sulfur batteries using TEGDME liquid electrolyte, Electrochim. Acta 52 (2006) 1563–1566.
[24] J.W. Choi, G. Cheruvally, D.S. Kim, J.H. Ahn, K.W. Kim, H.J. Ahn, Rechargeable lithium/sulfur battery with liquid electrolytes containing toluene as additive, J. Power Sources 183 (2008) 441–445.
[25] E. Peled, Y. Sternberg, A. Gorenshtein, Y. Lavi, Lithium-sulfur battery: evaluation of dioxolane-based electrolytes, J. Electrochem. Soc. 136 (1989) 1621–1625.
[26] D. Marmorstein, T.H. Yu, K.A. Striebel, F.R. McLarnon, J. Hou, E.J. Cairns, Electrochemical performance of lithium/sulfur cells with three different polymer electrolytes, J. Power Sources 89 (2000) 219–226, https://doi.org/10.1016/S0378-7753(00)00432-8.
[27] J. Shim, K.A. Striebel, E.J. Cairns, The lithium/sulfur rechargeable cell, J. Electrochem. Soc. 149 (2002)A1321https://doi.org/10.1149/1.1503076.
[28] C. Barchasz, J.C. Lepretre, S. Patoux, F. Alloin, Electrochemical properties of ether-based electrolytes for lithium/sulfur rechargeable batteries, Electrochim. Acta 89 (2013) 737–743, https://doi.org/10.1016/j.electacta.2012.11.001.
[29] J.-W. Choi, J.-K. Kim, G. Cheruvally, J.-H. Ahn, H.-J. Ahn, K.-W. Kim, Rechargeable lithium/sulfur battery with suitable mixed liquid electrolytes, Electrochim. Acta 52 (2007) 2075–2082, https://doi.org/10.1016/j.electacta.2006.08.016.
[30] D. Kodama, M. Kanakubo, M. Kokubo, S. Hashimoto, H. Nanjo, M. Kato, Density, viscosity, and solubility of carbon dioxide in glymes, Fluid Phase Equilib. 302 (2011) 103–108, https://doi.org/10.1016/j.fluid.2010.08.014.
[31] C. Barchasz, J.C. Lepretre, S. Patoux, F. Alloin, Revisiting TEGDME/DIOX binary electrolytes for lithium/sulfur batteries: importance of solvation ability and additives, J. Electrochem. Soc. 160 (2013) A430–A436, https://doi.org/10.1149/2.022303jes.
[32] Y. Mikhaylik, I. Kovalev, R. Schock, K. Kumaresan, J. Xu, J. Affinito, High energy rechargeable Li-S cells for EV application. status, remaining problems and solutions, ECS Trans. 25 (2010) 23–34, https://doi.org/10.1149/1.3414001.
[33] Y. Nimon, S.J. Visco, M.-Y. Chu, Dioxolane as a proctector for lithium electrodes, 2001. US patent 6, 225, 002.
[34] W.K. Wang, Y. Wang, Y.Q. Huang, C.J. Huang, Z.B. Yu, H. Zhang, et al., The electrochemical performance of lithium-sulfur batteries with $LiClO_4$ DOL/DME electrolyte, J. Appl. Electrochem. 40 (2010) 321–325, https://doi.org/10.1007/s10800-009-9978-z.
[35] T.J. Kim, B.O. Jeong, J.Y. Koh, S. Kim, Y. Jung, Influence of electrolyte composition on electrochemical performance of Li-S cells, Bull. Kor. Chem. Soc. 35 (2014) 1299–1304, https://doi.org/10.5012/bkcs.2014.35.5.1299.
[36] S. Kim, Y.J. Jung, H.S. Lim, The effect of solvent component on the discharge performance of lithium-sulfur cell containing various organic electrolytes, Electrochim. Acta 50 (2004) 889–892, https://doi.org/10.1016/j.electacta.2004.01.093.
[37] B. Jin, J.U. Kim, H.B. Gu, Electrochemical properties of lithium–sulfur batteries, J. Power Sources 117 (2003) 148–152.
[38] D.R. Chang, S.H. Lee, S.W. Kim, H.T. Kim, Binary electrolyte based on tetra(ethylene glycol) dimethyl ether and 1,3-dioxolane for lithium–sulfur battery, J. Power Sources 112 (2002) 452–460.

[39] S.E. Cheon, K.S. Ko, J.H. Cho, S.W. Kim, E.Y. Chin, H.T. Kim, Rechargeable lithium sulfur battery. I. Structural change of sulfur cathode during discharge and charge, J. Electrochem. Soc. 150 (2003) A796–A799.
[40] M.S. Song, S.C. Han, H.S. Kim, J.H. Kim, K.T. Kim, Y.M. Kang, H.J. Ahn, S.X. Dou, J. Y. Lee, Effects of nanosized adsorbing material on electrochemical properties of sulfur cathodes for Li/S secondary batteries, J. Electrochem. Soc. 151 (2004) A791–A795.
[41] Y.J. Jung, S. Kim, New approaches to improve cycle life characteristics of lithium–sulfur cells, Electrochem. Commun. 9 (2007) 249–254.
[42] K. Xu, Nonaqueous liquid electrolytes for lithium-based rechargeable batteries, Chem. Rev. 104 (10) (2004) 4303–4418.
[43] S. Zhang, K. Ueno, K. Dokko, M. Watanabe, Recent advances in electrolytes for lithium–sulfur batteries, Adv. Energy Mater. 5 (16) (2015). Special Issue: Understanding the Lithium–Sulfur Battery System at Relevant Scales, 1500117.
[44] J. Zhu, Advanced Separator Selection and Design for High-Performance Lithium-Sulfur Batteries, (Ph.D. dissertation), North Carolina State University, Raleigh, North Carolina, 2016.
[45] M. Balasubramanian, H.S. Lee, X. Sun, X.Q. Yang, A.R. Moodenbaugh, J. McBreen, D.A. Fischer, Z. Fu, Formation of SEI on cycled lithium-ion battery cathodes: soft X-ray absorption study, Electrochem. Solid-State Lett. 5 (1) (2002) A22–A25.
[46] J. Foropoulos, D.D. Desmarteau, Synthesis, properties, and reactions of bis ((trifluoromethyl)sulfonyl) imide, (Cf3so2)2nh, Inorg. Chem. 23 (1984) 3720–3723, https://doi.org/10.1021/Ic00191a011.
[47] L.J. Krause, W. Lamanna, J. Summerfield, M. Engle, G. Korba, R. Loch, et al., Corrosion of aluminum at high voltages in non-aqueous electrolytes containing perfluoroalkylsulfonyl imides; new lithium salts for lithium-ion cells, J. Power Sources 68 (1997) 320–325, https://doi.org/10.1016/S0378-7753(97)02517-2.
[48] A. Guerfi, S. Duchesne, Y. Kobayashi, A. Vijh, K. Zaghib, $LiFePO_4$ and graphite electrodes with ionic liquids based on bis(fluorosulfonyl)imide (FSI)(-) for Li-ion batteries, J. Power Sources 175 (2008) 866–873, https://doi.org/10.1016/j.jpowsour.2007.09.030.
[49] J.T. Dudley, D.P. Wilkinson, G. Thomas, R. Levae, S. Woo, H. Blom, et al., Conductivity of electrolytes for rechargeable lithium batteries, J. Power Sources 35 (1991) 59–82, https://doi.org/10.1016/0378-7753(91)80004-H.
[50] K. Xu, C.A. Angell, Sulfone-based electrolytes for lithium-ion batteries, J. Electrochem. Soc. 149 (2002) A920–A926, https://doi.org/10.1149/1.1483866.
[51] J.M. Tarascon, D. Guyomard, New electrolyte compositions stable over the O-V to 5-V voltage range and compatible with the Li1+Xmn2o4 carbon Li-ion cells, Solid State Ionics 69 (1994) 293–305, https://doi.org/10.1016/0167-2738(94)90418-9.
[52] M. Armand, The history of polymer electrolytes, Solid State Ionics 69 (1994) 309–319, https://doi.org/10.1016/0167-2738(94)90419-7.
[53] L.X. Yuan, J.K. Feng, X.P. Ai, Y.L. Cao, S.L. Chen, H.X. Yang, Improved dischargeability and reversibility of sulfur cathode in a novel ionic liquid electrolyte, Electrochem. Commun. 8 (2006) 610–614.
[54] J. Wang, S.Y. Chew, Z.W. Zhao, S. Ashraf, D. Wexler, J. Chen, S.H. Ng, S.L. Chou, H.K. Liu, Sulfur–mesoporous carbon composites in conjunction with a novel ionic liquid electrolyte for lithium rechargeable batteries, Carbon 46 (2008) 229–235.
[55] J.H. Shin, E.J. Cairns, N-Methyl-(n-butyl)pyrrolidinium bis (trifluoromethanesulfonyl)imide-LiTFSI–poly(ethylene glycol) dimethyl ether mixture as a Li/S cell electrolyte, J. Power Sources 177 (2008) 537–545.
[56] M. Armand, F. Endres, D.R. Macfarlane, H. Ohno, B. Scrosati, Ionic-liquid materials for the electrochemical challenges of the future, Nat. Mater. 8 (2009) 621–629, https://doi.org/10.1038/nmat2448.

[57] L.X. Yuan, J.K. Feng, X.P. Ai, Y.L. Cao, S.L. Chen, H.X. Yang, Improved discharge ability and reversibility of sulfur cathode in a novel ionic liquid electrolyte, Electrochem. Commun. 8 (2006) 610–614, https://doi.org/10.1016/j.elecom.2006. 02.007.
[58] J.-W. Park, K. Yamauchi, E. Takashima, N. Tachikawa, K. Ueno, K. Dokko, et al., Solvent effect of room temperature ionic liquids on electro- chemical reactions in lithium-sulfur batteries, J. Phys. Chem. C 117 (2013) 4431–4440, https://doi.org/10.1021/jp400153m.
[59] J.W. Park, K. Ueno, N. Tachikawa, K. Dokko, M. Watanabe, Ionic liquid electrolytes for lithium-sulfur batteries, J. Phys. Chem. C 117 (2013) 20531–20541, https://doi.org/10.1021/Jp408037e.
[60] N. Tachikawa, K. Yamauchi, E. Takashima, J.W. Park, K. Dokko, M. Watanabe, Reversibility of electrochemical reactions of sulfur supported on inverse opal carbon in glyme-Li salt molten complex electrolytes, Chem. Commun. (Camb.) 47 (2011) 8157–8159, https://doi.org/10.1039/c1cc12415c.
[61] Y. Yan, Y.X. Yin, S. Xin, J. Su, Y.G. Guo, L.J. Wan, High-safety lithium-sulfur battery with prelithiated Si/C anode and ionic liquid electrolyte, Electrochim. Acta 91 (2013) 58–61, https://doi.org/10.1016/j.electacta.2012.12.077.
[62] S. Kim, Y. Jung, S.J. Park, Effect of imidazolium cation on cycle life characteristics of secondary lithium-sulfur cells using liquid electrolytes, Electrochim. Acta 52 (2007) 2116–2122, https://doi.org/10.1016/j.electacta.2006.08.028.
[63] J.H. Shin, E.J. Cairns, Characterization of N-methyl-N-butylpyrrolidinium Bis (trifluoromethanesulfonyl)imide-LiTFSI-Tetra(ethylene glycol) dimethyl ether mixtures as a Li metal cell electrolyte, J. Electrochem. Soc. 155 (2008) A368–A373, https://doi.org/10.1149/1.2869876.
[64] K. Dokko, N. Tachikawa, K. Yamauchi, M. Tsuchiya, A. Yamazaki, E. Takashima, et al., Solvate ionic liquid electrolyte for Li-S batteries, J. Electrochem. Soc. 160 (2013) A1304–A1310, https://doi.org/10.1149/2.111308jes.
[65] L.N. Wang, H.R. Byon, N-Methyl-N-propylpiperidinium bis(trifluoromethanesulfonyl) imide-based organic electrolyte for high performance lithium-sulfur batteries, J. Power Sources 236 (2013) 207–214, https://doi.org/10.1016/j.jpowsour.2013.02.068.
[66] T. Yim, M.-S. Park, J.-S. Yu, K.J. Kim, K.Y. Im, J.-H. Kim, et al., Effect of chemical reactivity of polysulfide toward carbonate-based electrolyte on the electrochemical performance of Li-S batteries, Electrochim. Acta 107 (2013) 454–460, https://doi.org/10.1016/j.electacta.2013.06.039.
[67] J. Wang, Y.W. Wang, X.M. He, J.G. Ren, C.Y. Jiang, C.R. Wan, J. Power Sources 138 (2004) 271–273.
[68] J.L. Wang, J. Yang, J.Y. Xie, N.X. Xu, Adv. Mater. 14 (2002) 963–996.
[69] J.L. Wang, J. Yang, C.R. Wan, K. Du, J.Y. Xie, N.X. Xu, Adv. Funct. Mater. 13 (2003) 487–492.
[70] J.L. Wang, L. Liu, Z.J. Ling, J. Yang, C.R. Wan, C.Y. Jiang, Electrochim. Acta 48 (2003) 1861–1867.
[71] J. Hassoun, B. Scrosati, Angew. Chem. Int. Ed. 49 (2010) 2371–2374.
[72] J. Hassoun, Y.-K. Sun, B. Scrosati, J. Power Sources 196 (2011) 343–348.
[73] B. Zhang, C. Lai, Z. Zhou, X.P. Gao, Electrochim. Acta 54 (2009) 3708–3713.
[74] C. Lai, X.P. Gao, B. Zhang, T.Y. Yan, Z. Zhou, J. Phys. Chem. C 113 (2009) 4712–4716.
[75] B. Zhang, X. Qin, G.R. Li, X.P. Gao, Energy Environ. Sci. 3 (2010) 1531–1537.
[76] S. Xin, L. Gu, N.H. Zhao, Y.X. Yin, L.J. Zhou, Y.G. Guo, et al., Smaller sulfur molecules promise better lithium-sulfur batteries, J. Am. Chem. Soc. 134 (2012) 18510–18513, https://doi.org/10.1021/ja308170k.

[77] Z. Li, L. Yuan, Z. Yi, Y. Sun, Y. Liu, Y. Jiang, et al., Insight into the electrode mechanism in lithium-sulfur batteries with ordered microporous carbon confined sulfur as the cathode, Adv. Energy Mater. 4 (2013) 473–477, https://doi.org/10.1002/aenm.201301473.
[78] J.L. Wang, J. Yang, J.Y. Xie, N.X. Xu, Y. Li, Sulfur-carbon nano-composite as cathode for rechargeable lithium battery based on gel electrolyte, Electrochem. Commun. 4 (2002) 499–502, https://doi.org/10.1149/1.1461377.
[79] D.W. Wang, G. Zhou, F. Li, K.H. Wu, G.Q. Lu, H.M. Cheng, et al., A microporous-mesoporous carbon with graphitic structure for a high-rate stable sulfur cathode in carbonate solvent-based Li-S batteries, Phys. Chem. Chem. Phys. 14 (2012) 8703–8710, https://doi.org/10.1039/c2cp40808b.
[80] S. Zheng, P. Han, Z. Han, H. Zhang, Z. Tang, J. Yang, High performance C/S composite cathodes with conventional carbonate-based electrolytes in Li-S battery, Sci. Rep. 4 (2014)4842https://doi.org/10.1038/srep04842.
[81] Z. Liu, W. Fu, C. Liang, Chapter 24 Lithium–sulfur batteries, in: C. Daniel, J.O. Besenhard (Eds.), Handbook of Battery Materials, second ed., Wiley-VCH Verlag GmbH & Co. KGaA, 2011.
[82] M. Liu, X. Qin, Y.-B. He, B. Li, F. Kang, Recent innovative configurations in high-energy lithium–sulfur batteries, Mater. Chem. A 5 (2017) 5222–5234.
[83] S. Jeong, Y. Lim, Y. Choi, G. Cho, K. Kim, H. Ahn, K. Cho, Electrochemical properties of lithium sulfur cells using PEO polymer electrolytes prepared under three different mixing conditions, J. Power Sources 174 (2007) 745–750.
[84] Y.M. Lee, N. Choi, J.H. Park, J. Park, Electrochemical performance of lithium/sulfur batteries with protected Li anodes, J. Power Sources 119 (2003) 964–972.
[85] M. Rao, X. Geng, X. Li, S. Hu, W. Li, Lithium-sulfur cell with combining carbon nanofibers–sulfur cathode and gel polymer electrolyte, J. Power Sources 212 (2012) 179–185.
[86] M. Nagao, A. Hayashi, M. Tatsumisago, Sulfur–carbon composite electrode for all-solid-state Li/S battery with $Li_2S–P_2S_5$ solid electrolyte, Electrochim. Acta 56 (2011) 6055–6059.
[87] A. Hayashi, T. Ohtomo, F. Mizuno, K. Tadanaga, M. Tatsumisago, All-solid-state Li/S batteries with highly conductive glass–ceramic electrolytes, Electrochem. Commun. 5 (2003) 701–705.
[88] A. Unemoto, S. Yasaku, G. Nogami, M. Tazawa, M. Taniguchi, M. Matsuo, T. Ikeshoji, S. Orimo, Development of bulk-type all-solid-state lithium-sulfur battery using LiBH4 electrolyte, Appl. Phys. Lett. 105 (2014)083901.
[89] S. Wang, Y. Ding, G. Zhou, G. Yu, A. Manthiram, Durability of the $Li_{1+x}Ti_{2-x}Al_x(PO_4)_3$ solid electrolyte in lithium–sulfur batteries, ACS Energy Lett. 1 (2016) 1080–1085.
[90] F. Han, J. Yue, X. Fan, T. Gao, C. Luo, Z. Ma, L. Suo, C. Wang, High-performance all-solid-state lithium–sulfur battery enabled by a mixed-conductive Li2S nanocomposite, Nano Lett. 16 (2016) 4521–4527.
[91] M. Nagao, Y. Imade, H. Narisawa, T. Kobayashi, R. Watanabe, T. Yokoi, T. Tatsumi, R. Kanno, All-solid-state Li–sulfur batteries with mesoporous electrode and thio-LISICON solid electrolyte, J. Power Sources 222 (2013) 237–242.
[92] A. Manuel Stephan, Review on gel polimer electrolytes for lithium batteries, Eur. Polym. J. 42 (2006) 21–42, https://doi.org/10.1016/j.eurpolymj.2005.09.017.
[93] D.E. Fenton, J.M. Parker, P.V. Wright, Complexes of alkali-metalions with poly (ethyleneoxide), Polymer 14 (1973) 589, https://doi.org/10.1016/0032-3861(73)90146-8.

[94] N.H. Idris, M.M. Rahman, J.Z. Wang, H.K. Liu, Microporous gel polimer electrolytes for lithium rechargeable battery application, J. Power Sources 201 (2012) 294–300, https://doi.org/10.1016/j.jpowsour.2011.10.141.
[95] J.Y. Song, Y.Y. Wang, C.C. Wan, Review of gel-type polimer electrolytes for lithium-ion batteries, J. Power Sources 77 (1999) 183–197, https://doi.org/10.1016/S0378-7753(98)00193-1.
[96] K.M. Kim, N.G. Park, K.S. Ryu, S.H. Chang, Characteristics of PVdF-HFP/TiO_2 composite membrane electrolytes prepared by phase inversion and conventional casting methods, Electrochim. Acta 51 (2006) 5636–5644, https://doi.org/10.1016/j.electacta.2006.02.038.
[97] Y. Zhao, Y. Zhang, D. Gosselink, T.N. Doan, M. Sadhu, H.J. Cheang, et al., Polymer electrolytes for lithium/sulfur batteries, Membranes (Basel, Switz.) 2 (2012) 553–564, https://doi.org/10.3390/membranes2030553.
[98] J.E. Weston, B.C.H. Steele, Effects of inert fillers on the mechanical and electrochemical properties of lithium salt poly(ethylene-oxide) polymer electrolytes, Solid State Ionics 7 (1982) 75–79, https://doi.org/10.1016/0167-2738(82)90072-8.
[99] M.-Y. Chu, Rechargeable positive electrodes, 1998. WIPO Patent Application WO/1998/005085.
[100] B.H. Jeon, J.H. Yeon, K.M. Kim, I.J. Chung, Preparation and electrochemical properties of lithium–sulfur polymer batteries, J. Power Sources 109 (2002) 89–97.
[101] B.H. Jeon, J.H. Yeon, I.J. Chung, Preparation and electrical properties of lithium-sulfur-composite polymer batteries, J. Mater. Process. Technol. 143 (2003) 93–97, https://doi.org/10.1016/S0924-0136(03)00327-3.
[102] G.B. Appetecchi, F. Croce, L. Persi, F. Ronci, B. Scrosati, Transport and interfacial properties of composite polymer electrolytes, Electrochim. Acta 45 (2000) 1481–1490, https://doi.org/10.1016/S0013-4686(99)00363-1.
[103] S.H. Chung, Y. Wang, L. Persi, F. Croce, S.G. Greenbaum, B. Scrosati, et al., Enhancement of ion transport in polymer electrolytes by addition of nanoscale inorganic oxides, J. Power Sources 9 (2001) 644–648, https://doi.org/10.1016/S0378-7753(01)00748-0.
[104] F. Croce, L. Persi, B. Scrosati, F. Serraino-Fiory, E. Plichta, M.A. Hendrickson, Role of the ceramic fillers in enhancing the transport properties of composite polymer electrolytes, Electrochim. Acta 46 (2001) 2457–2461, https://doi.org/10.1016/S0013-4686(01)00458-3.
[105] J.H. Shin, K.W. Kim, H.J. Ahn, J.H. Ahn, Electrochemical properties and interfacial stability of (PEO) (10)LiCF3SO3-TinO2n-1 composite polymer electrolytes for lithium/sulphur battery, Mater. Sci. Eng. B Solid State Mater. Adv. Technol. 95 (2002) 148–156, https://doi.org/10.1016/S0921-5107(02)00226-X.
[106] J.H. Ahn, G.X. Wang, H.K. Liu, S.X. Dou, Nanoparticle-dispersed PEO polimer electrolytes for Li batteries, J. Power Sources 119 (2003) 422–426, https://doi.org/10.1016/S0378-7753(03)00264-7.
[107] M.A.K.L. Dissanayake, P.A.R.D. Jayathilaka, R.S.P. Bokalawala, I. Albinsson, B.-E. Mellander, Effect of concentration and grain size of alumina filler on the ionic conductivity enhancement of the (PEO) 9LiCF3SO3:Al2O3 composite polymer electrolyte, J. Power Sources 119–121 (2003) 409–414, https://doi.org/10.1016/s0378-7753(03)00262-3.
[108] C.W. Lin, C.L. Hung, M. Venkateswarlu, B.J. Hwang, Influence of TiO2 nanoparticles on the transport properties of composite polymerel ectrolyte for lithium-ion batteries, J. Power Sources 146 (2005) 397–401, https://doi.org/10.1016/j.jpowsour.2005.03.028.

[109] X.J. Zhu, Z.Y. Wen, Z.H. Gu, Z.X. Lin, Electrochemical characterization and performance improvement of lithium/sulphur polymer batteries, J. Power Sources 139 (2005) 269–273, https://doi.org/10.1016/j.jpowsour.2004.07.002.
[110] S.S. Jeong, Y.T. Lim, Y.J. Choi, G.B. Cho, K.W. Kim, H.J. Ahn, et al., Electrochemical properties of lithium sulphur cells using PEO polymer electrolytes prepared under three different mixing conditions, J. Power Sources 174 (2007) 745–750, https://doi.org/10.1016/j.jpowsour.2007.06.108.
[111] J. Shin, B. Jung, S. Jeong, K. Kim, H. Ahn, K. Cho, J. Ahn, Electrochemical and interfacial properties of $(PEO)_{10}LiCF_3SO_3-Al_2O_3$ nanocomposite polymer electrolytes using ball milling, Met. Mater. Int. 10 (2004) 177–183.
[112] J.H. Shin, K.W. Kim, H.J. Ahn, J.H. Ahn, Electrochemical properties and interfacial stability of $(PEO)_{10}LiCF_3SO_3$–Ti_nO_{2n-1} composite polymer electrolytes for lithium/sulfur battery, Mater. Sci. Eng. B Solid State Mater. Adv. Technol. 95 (2002) 148–156.
[113] X.J. Zhu, Z.Y. Wen, Z.H. Gu, Z.X. Lin, Electrochemical characterization and performance improvement of lithium/sulfur polymer batteries, J. Power Sources 139 (2005) 269–273.
[114] J. Hassoun, B. Scrosati, Moving to a solid-state configuration: a valid approach to making lithium-sulfur batteries viable for practical applications, Adv. Mater. 22 (2010) 5198–5201.
[115] J. Przyluski, M. Siekierski, W. Wieczorek, Effective-medium theory in studies of conductivity of composite polymeric electrolytes, Electrochim. Acta 40 (1995) 2101–2108, https://doi.org/10.1016/0013-4686(95)00147-7.
[116] F. Croce, S. Sacchetti, B. Scrosati, Advanced, high-performance composite polymer electrolytes for lithium batteries, J. Power Sources 161 (2006) 560–564, https://doi.org/10.1016/j.jpowsour.2006.03.069.
[117] F. Croce, L. Settimi, B. Scrosati, Super acid ZrO2-added, composite polymer electrolytes with improved transport properties, Electrochem. Commun. 8 (2006) 364–368, https://doi.org/10.1016/j.elecom.2005.12.002.
[118] X. Liang, Z. Wen, Y. Liu, H. Zhang, L. Huang, J. Jin, Highly dispersed sulphur in ordered mesoporous carbon sphere as a composite cathode for rechargeable polymer Li/S battery, J. Power Sources 196 (2011) 3655–3658, https://doi.org/10.1016/j.jpowsour.2010.12.052.
[119] J.J. Xu, H. Ye, Polymer gel electrolytes based on oligomeric polyether/cross-linked PMMA blends prepared via in situ polymerization, Electrochem. Commun. 7 (2005) 829–835, https://doi.org/10.1016/j.elecom.2005.04.034.
[120] G.C. Li, Z.H. Li, P. Zhang, H.P. Zhang, Y.P. Wu, Research on a gel polymer electrolyte for Li-ion batteries, Pure Appl. Chem. 80 (2008) 2553–2563, https://doi.org/10.1351/pac200880112553.
[121] J.H. Shin, S.S. Jung, K.W. Kim, H.J. Ahn, J.H. Ahn, Preparation and characterization of plasticized polimer electrolytes based on the PVdF-HFP copolymer for lithium/sulphur battery, J. Mater. Sci. Mater. Electron. 13 (2002) 727–733, https://doi.org/10.1023/A:1021521207247.
[122] H.-S. Ryu, H.-J. Ahn, K.-W. Kim, J.-H. Ahn, J.-Y. Lee, Discharge process of Li/PVdF/S cells at room temperature, J. Power Sources 153 (2006) 360–364, https://doi.org/10.1016/j.jpowsour.2005.05.037.
[123] S.S. Zhang, D.T. Tran, How a gel polymer electrolyte affects performance of lithium/sulphur batteries, Electrochim. Acta 114 (2013) 296–302, https://doi.org/10.1016/j.electacta.2013.10.069.
[124] J.W. Choi, J.H. Kim, G. Cheruvally, J.H. Ahn, K.W. Kim, H.J. Ahn, J.U. Kim, Microporous poly(vinylidene fluoride-co-hexafluoropropylene) polymer electrolytes for lithium/sulfur cells, J. Ind. Eng. Chem. 12 (2006) 939–949.

[125] Z.W. Seh, Q. Zhang, W. Li, G. Zheng, H. Yao, Y. Cui, Chem. Sci. 4 (2013) 3673–3677.
[126] K. Park, J.H. Cho, J. Jang, B. Yu, T. Andreah, K.M. Miller, C.J. Ellison, J.B. Goodenough, Energy Environ. Sci. 8 (2015) 2389–2395.
[127] M. Liu, D. Zhou, Y. He, Y. Fu, X. Qin, C. Miao, H. Du, B. Li, Q. Yang, Z. Lin, Nano Energy 22 (2016) 278–289.
[128] M. Liu, H. Jiang, Y. Ren, D. Zhou, F. Kang, T. Zhao, Electrochim. Acta 213 (2016) 871–878.
[129] G.Y. Adachi, N. Imanaka, H. Aono, Fast Li-circle plus conducting ceramic electrolytes, Adv. Mater. 8 (1996) 127–135, https://doi.org/10.1002/adma.19960080205.
[130] A.D. Robertson, A.R. West, A.G. Ritchie, Review of crystalline lithium-ion conductors suitable for high temperature battery applications, Solid State Ionics 104 (1997) 1–11, https://doi.org/10.1016/S0167-2738(97)00429-3.
[131] V. Thangadurai, W. Weppner, Recent progress in solid oxide and lithium ion conducting electrolytes research, Ionics 12 (2006) 81–92, https://doi.org/10.1007/s11581-006-0013-7.
[132] V. Thangadurai, S. Narayanan, D. Pinzaru, Garnet-type solid-state fast Li ion conductors for Li batteries: critical review, Chem. Soc. Rev. 43 (2014) 4714–4727, https://doi.org/10.1039/c4cs00020j.
[133] S. Stramare, V. Thangadurai, W. Weppner, Lithium lanthanum titanates: a review, Chem. Mater. 15 (2003) 3974–3990, https://doi.org/10.1021/Cm0300516.
[134] T. Kobayashi, Y. Imade, D. Shishihara, K. Homma, M. Nagao, R. Watanabe, et al., All solid-state battery with sulfur electrode and thio-LISICON electrolyte, J. Power Sources 182 (2008) 621–625, https://doi.org/10.1016/j.jpowsour.2008.03.030.
[135] N. Kamaya, K. Homma, Y. Yamakawa, M. Hirayama, R. Kanno, M. Yonemura, T. Kamiyama, Y. Kato, S. Hama, K. Kawamoto, A. Mitsui, A lithium superionic conductor, Nat. Mater. 10 (2011) 682–686, https://doi.org/10.1038/nmat3066.
[136] A. Hayashi, S. Hama, H. Morimoto, M. Tatsumisago, T. Minami, Preparation of Li_2S-P_2S_5 amorphous solid electrolytes by mechanical milling, J. Am. Ceram. Soc. 84 (2001) 477–479, https://doi.org/10.1111/j.1151-2916.2001.tb00685.x.
[137] A. Hayashi, T. Ohtomo, F. Mizuno, K. Tadanaga, M. Tatsumisago, Rechargeable lithium batteries, using sulfur-based cathode materials and Li2S-P2S5 glass-ceramic electrolytes, Electrochim. Acta 50 (2004) 893–897, https://doi.org/10.1016/j.electacta.2004.02.061.
[138] A. Hayashi, R. Ohtsubo, T. Ohtomo, F. Mizuno, M. Tatsumisago, All-solid-state rechargeable lithium batteries with Li_2S as a positive electrode material, J. Power Sources 183 (2008) 422–426.
[139] M. Nagao, A. Hayashi, M. Tatsumisago, Fabrication of favorable interface between sulfide solid electrolyte and Li metal electrode for bulk-type solid-state Li/S battery, Electrochem. Commun. 22 (2012) 177–180, https://doi.org/10.1016/j.elecom.2012.06.015.
[140] S. Teragawa, K. Aso, K. Tadanaga, A. Hayashi, M. Tatsumisago, Preparation of Li_2S-P_2S_5 solid electrolyte from N-methylformamide solution and application for all-solid-state lithium battery, J. Power Sources 248 (2014) 939–942, https://doi.org/10.1016/j.jpowsour.2013.09.117.
[141] M. Agostini, Y. Aihara, T. Yamada, B. Scrosati, J. Hassoun, A lithium–sulfur battery using a solid, glass-type P_2S_5–Li_2S electrolyte, Solid State Ionics 244 (2013) 48–51.
[142] T. Yamada, S. Ito, R. Omoda, T. Watanabe, Y. Aihara, M. Agostini, U. Ulissi, J. Hassoun, B. Scrosati, All solid-state lithium–sulfur battery using a glass-type P_2S_5–Li_2S electrolyte: benefits on anode kinetics, J. Electrochem. Soc. 162 (2015) A646–A651.

[143] A. Hayashi, S. Hama, H. Morimoto, M. Tatsumisago, T. Minami, High lithium ion conductivity of glass-ceramics derived from mechanically milled glassy powders, Chem. Lett. 84 (2001) 872–873, https://doi.org/10.1246/cl.2001.872.
[144] A. Hayashi, S. Hama, T. Minami, M. Tatsumisago, Formation of superionic crystals from mechanically milled Li_2S-P_2S_5 glasses, Electrochem. Commun. 5 (2003) 111–114, https://doi.org/10.1016/S1388-2481(02)00555-6.
[145] H. Nagata, Y. Chikusa, A lithium sulfur battery with high power density, J. Power Sources 264 (2014) 206–210, https://doi.org/10.1016/j.jpowsour.2014.04.106.
[146] Z.C. Liu, W.J. Fu, E.A. Payzant, X. Yu, Z.L. Wu, N.J. Dudney, et al., Anomalous high ionic conductivity of nanoporous beta-Li_3PS_4, J. Am. Chem. Soc. 135 (2013) 975–978, https://doi.org/10.1021/Ja3110895.
[147] Z. Lin, Z.C. Liu, N.J. Dudney, C.D. Liang, Lithium superionic sulfide cathode for all-solid lithium-sulfur batteries, ACS Nano 7 (2013) 2829–2833, https://doi.org/10.1021/Nn400391h.
[148] Z. Lin, Z.C. Liu, W.J. Fu, N.J. Dudney, C.D. Liang, Lithium polysulfidophosphates: a family of lithium-conducting sulfur-rich compounds for lithium-sulfur batteries, Angew. Chem. Int. Ed. 52 (2013) 7460–7463, https://doi.org/10.1002/anie.201300680.
[149] K. Minami, F. Mizuno, A. Hayashi, M. Tatsumisago, Lithium ion conductivity of the Li_2S–P_2S_5 glass-based electrolytes prepared by the melt, Solid State Ionics 178 (2007) 837–841.
[150] N. Machida, K. Kobayashi, Y. Nishikawa, T. Shigematsu, Electrochemical properties of sulfur as cathode materials in a solid-state lithium battery with inorganic solid electrolytes, Solid State Ionics 175 (2004) 247–250.
[151] N. Machida, T. Shigematsu, An all-solid-state lithium battery with sulfur as positive electrode materials, Chem. Lett. 33 (2004) 376–377.
[152] Z. Liu, F. Huang, J. Yang, B. Wang, J. Sun, New lithium ion conductor, thio-LISICON lithium zirconium sulfide system, Solid State Ionics 179 (2008) 1714–1716.
[153] Z. Liu, F. Huang, J. Yang, Y. Wang, J. Sun, Preparation of new lithium ion composite electrolyte $3Li_4SiS_4$–$0.5La_2S_3$ by mechanical milling, Solid State Sci. 10 (2008) 1429–1433.
[154] N. Azimi, Z. Xue, I. Bloom, M.L. Gordin, D. Wang, T. Daniel, C. Takoudis, Z. Zhang, Understanding the effect of a fluorinated ether on 186 the performance of lithium–sulfur batteries, ACS Appl. Mater. Interfaces 7 (17) (2015) 9169–9177.
[155] N. Azimi, Z. Xue, N.D. Rago, C. Takoudis, M.L. Gordin, J. Song, D. Wang, Z. Zhang, Fluorinated electrolytes for Li-S battery: suppressing the self-discharge with an electrolyte containing fluoroether solvent, J. Electrochem. Soc. 162 (1) (2015) A64–A68.
[156] S.S. Zhang, New insight into liquid electrolyte of rechargeable lithium/sulfur battery, Electrochim. Acta 97 (2013) 226–230.
[157] D. Aurbach, E. Pollak, R. Elazari, G. Salitra, C.S. Kelley, J. Affinito, On the surface chemical aspects of very high energy density, rechargeable Li-sulfur batteries, J. Electrochem. Soc. 156 (2009) A694, https://doi.org/10.1149/1.3148721.
[158] S.S. Zhang, Role of $LiNO_3$ in rechargeable lithium/sulfur battery, Electrochim. Acta 70 (2012) 344–348, https://doi.org/10.1016/j.electacta.2012.03.081.
[159] S. Xiong, K. Xie, Y. Diao, X. Hong, Characterization of the solid electrolyte interphase on lithium anode for preventing the shuttle mechanism in lithium–sulfur batteries, J. Power Sources 246 (2014) 840–845.
[160] M. Kazazi, M.R. Vaezi, A. Kazemzadeh, Improving the self-discharge behavior of sulfur-polypyrrole cathode material by $LiNO_3$ electrolyte additive, Ionics 20 (9) (2014) 1291–1300.

[161] R. Demir-Cakan, M. Morcrette, B. Gangulibabu, A. Guéguen, R. Dedryvère, J.-M. Tarascon, Li-S batteries: simple approaches for superior performance, Energy Environ. Sci. 6 (2013) 176–182, https://doi.org/10.1039/c2ee23411d.
[162] S.Z. Xiong, K. Xie, Y. Diao, X.B. Hong, On the role of polysulfides for a stable solid electrolyte interphase on the lithium anode cycled in lithium-sulfur batteries, J. Power Sources 236 (2013) 181–187, https://doi.org/10.1016/j.jpowsour.2013.02.072.
[163] C. Barchasz, F. Mesguich, J. Dijon, J.C. Lepretre, S. Patoux, F. Alloin, Novel positive electrode architecture for rechargeable lithium/sulfur batteries, J. Power Sources 211 (2012) 19–26, https://doi.org/10.1016/j.jpowsour.2012.03.062.
[164] S.S. Zhang, J.A. Read, A new direction for the performance improvement of rechargeable lithium/sulfur batteries, J. Power Sources 200 (2012) 77–82, https://doi.org/10.1016/j.jpowsour.2011.10.076.
[165] S.S. Zhang, D.T. Tran, A proof-of-concept lithium/sulfur liquid battery with exceptionally high capacity density, J. Power Sources 211 (2012) 169–172, https://doi.org/10.1016/j.jpowsour.2012.04.006.
[166] Z. Lin, Z. Liu, W. Fu, N.J. Dudney, C. Liang, Phosphorous pentasulfide as a novel additive for high-performance lithium-sulfur batteries, Adv. Funct. Mater. 23 (2013) 1064–1069, https://doi.org/10.1002/adfm.201200696.
[167] A. Touidjine, Lithium sulfur batteries: mechanisms, modelling and materials conference. Recent advances in lithium-sulfur batteries research, Johnson Matthey Technol. Rev. 61 (4) (2017) 308–310.
[168] L. Lodovico, A. Varzi, S. Passerini, Radical decomposition of ether-based electrolytes for Li-S batteries, J. Electrochem. Soc. 164 (9) (2017) A1812–A1819, https://doi.org/10.1149/2.0311709jes.
[169] K. Kumai, T. Ikeya, K. Ishihara, T. Iwahori, N. Imanishi, Y. Takeda, O. Yamamoto, Degradation mechanism due to decomposition of organic electrolyte in Li/MoS2 cells during long cycling, J. Power Sources 70 (2) (1998) 235–239. ISSN: 0378-7753.
[170] J. Balach, T. Jaumann, L. Giebeler, Nanosized Li2S-based cathodes derived from MoS2 for high-energy density Li–S cells and Si–Li2S full cells in carbonate-based electrolyte, Energy Storage Mater. 8 (2017) 209–216.

CHAPTER EIGHT

Negative Electrode (Anode) Materials

The anode is an important part of the Li-S battery system, because the long-term stability of the Li-S battery cycle depends on the stability of the anode. Metallic lithium is widely used as an anode material for Li-S batteries due to its low potential and high capacity, resulting in higher energy density. Nevertheless, metallic lithium is unstable in organic electrolytes, which can pose problems related to battery safety.

The high reactivity of the anode lithium and the dendritic growth of lithium deposits make the practical use of Li-S batteries more difficult. According to [1], the side deposition and corrosion of polysulfides on the lithium anode make these issues even more challenging. Therefore, anode protection is critical for suppressing the intrinsic lithium dendrite and polysulfide corrosion and as a result improving the performances of Li-S batteries.

Within a few decades of lithium battery development, research on metallic lithium anodes achieved very limited success. In the Li-S system, the problems differ from those observed in lithium batteries with the use of transition metal oxides or phosphates as a cathode. The high mobility of sulfur in the electrochemical cell changes the behavior of the metallic lithium.

According to [2], metallic lithium electrodes have the potential to significantly increase the energy density. But the disadvantage of lithium in this form is the formation of its dendrites on the surface of the metal, which can degrade performance and even lead to short circuits.

Replacing metallic lithium with other high capacity nonlithium anode materials can enable the Li-S battery to resist a short circuit due to limited lithium ions [3, 4].

High capacity nonlithium anode materials, like boron [5], silicon [6] and tin [7], were paired with sulfur cathodes. However, the huge volumetric expansion of such anode materials during charge/discharge processes resulted in a great loss in active materials and a rapid decay rate during cycling [8–10].

Carbonate-based electrolytes are the most frequently used electrolytes for these anodes. However, ether-based electrolyte systems are better for Li-S batteries compared with carbonate-based electrolytes, because polysulfides

Next-Generation Batteries With Sulfur Cathodes
https://doi.org/10.1016/B978-0-12-816392-4.00008-6

react with carbonates. Therefore, a different anodic solid electrolyte interphase (SEI) is generated in ether-based electrolytes compared with that in carbonated-based electrolytes [9–14].

The shortage of compatible comprehension between the anode and ether-based electrolyte (especially different SEI formation mechanisms) results in a fast capacity decay and short cycle life for the current batteries [15]. Moreover, limited lithium ions are supplied in these lithium ion sulfur cells compared with Li-S batteries, which would be irreversibly lost in side reactions. When the capacity of the electrodes does not match, the deep discharge or overcharge of cathode or anode is detrimental to the cycling performance, and thus overcycling might occur since only the voltage of the full cell is controlled. Problems caused by lithium dendrites in Li-S batteries can be avoided in such metal-free batteries, while the superiority of high specific energy can still be maintained [4, 16, 17].

8.1 MECHANISMS OF FORMATION OF LITHIUM DENDRITES

Lithium dendrites formed by the heterogeneous deposition of lithium to the current collector cause short-circuit risks and capacity loss for batteries. The mechanism of formation of lithium dendrites is not yet fully recognized. A very detailed analysis of the formation of dendrites for lithium-ion batteries has recently been carried out.

In [18] it was reported that there are two types of sediments that consist of lithium fibers (Fig. 33).

Clustered "mossy" sediments, which arise at low speeds, grow from the roots and can be relatively easy to control. Rare and faster emerging "dendritic" performances, which grow only at their ends, are more difficult to control and are responsible for most of the problems.

The growth of mossy structures arising from roots can be blocked by adding a separator layer made of a nanoporous ceramic material. The increase in dendritic structures at the ends cannot be easily blocked, but luckily the normal operating currents for practical batteries are much lower than the current associated with tailings, so they will not be formed if the electrolyte degradation occurs.

Theoretically, replacing carbon anodes with those based on metallic lithium can reduce the weight and volume of lithium-ion batteries by half. But these dendrites are still poorly understood. These small fibers can pass straight through the separator and cause an explosion or fire.

Fig. 33 To the right, scanning electron microscope (SEM) images show two types of lithium deposits: a massive, mossy type (top) that grows from its base, and a needle-like dendritic type (bottom) that grows from the tip. On the left, SEM images indicate the blocking effect of the ceramic material layer, which limits the growth of mossy sediments [18].

Most of the earlier studies on the use of metallic lithium anodes were carried out at low current levels or low battery power, and thus the second type of growth mechanism was not reliably observed. Conducted tests at higher current levels clearly revealed two distinct types of growth.

The new findings have provided a better understanding of the basic scientific principles and demonstrated the speed and capacity limitations that are available.

Similar mechanisms of dendrite formation may also occur in the case of lithium-sulfur cells with solid lithium anodes.

Various electrolyte additives (such as $LiNO_3$ [19–21] and InI_3 [22]) and anodic protection interlayers (such as PEDOT-*co*-PEG [23] and Li_3N [24]) were utilized in Li-S cells to alleviate the dendrite growth. These additives can promote anodic SEI formation on the surface of Li metal to isolate the corrosion from polysulfides.

According to [25], the Li dendrite problem is exacerbated at high current densities. A regimen for dendrite evolution in which the reverse is true was reported. It was found that when the plating and stripping current density was raised above $\sim 9\,mA/cm^2$, there was substantial self-heating of the dendrites, which triggers extensive surface migration of Li. Such surface diffusion healed the dendrites and smoothed the Li metal surface. Repeated doses of high current-density healing treatment enabled the safe cycling of Li-S batteries with high coulombic efficiency.

8.2 LITHIUM ANODE

The metallic lithium has a large theoretical capacity of ~3800 $mAh\,g^{-1}$, the largest of all anode materials for lithium batteries. However, lithium degrades during cyclic operation, creating dendrites that negatively affect the efficiency of cyclic work and create safety problems for renewable batteries [26–29].

The formation of dendrites and the low efficiency of lithium cyclic work occur due to the instability of the SEI film on its surface, which cannot inhibit changes in the shape and volume of the lithium electrode during the loading/unloading process, and consequently the formation of lithium dendrites. Intermediate polysulfides in Li-S batteries can be dissolved in the organic electrolyte, which causes serious side reactions on the lithium surface, making it difficult to create a permanent SEI film. Some approaches have been used to solve this problem, including the use of an additive (e.g., $LiNO_3$, polysulfide, P_2S_5) in electrolytes.

An important feature of metallic lithium anodes in Li-S batteries is their freedom from dendritic growth during lithium deposition. Assigned to the phenomenon of polysulfide transfer, soluble polysulfide forms corrode the lithium anode. Dendrites cannot survive the reaction with polysulfides. However, because of corrosion of the metallic lithium anodes, the insoluble Li_2S_2 and/or Li_2S settle on the surface of the anode, passivating the lithium electrode. The passivation layer gradually increases during cyclic battery operation, which leads to two negative effects on Li-S batteries:

(i) a rapid drop in capacity due to the consumption of sulfur to create a passivation layer,

(ii) high cell resistance caused by the electronically and ionically insulating character of Li_2S [30].

According to [31], by controlling the concentration of polysulfides and $LiNO_3$ in the electrolyte, a stable and uniform SEI is formed on the lithium surface, which suppresses the side reactions and dendrite growth. It was proven in [32] that an electrolyte based on 0.02 M Li_2S_5 (0.10 M sulfur)–5.0 wt% $LiNO_3$ could improve the cycling stability of the lithium metal anode. However, the very high concentration polysulfides (>0.5M) in

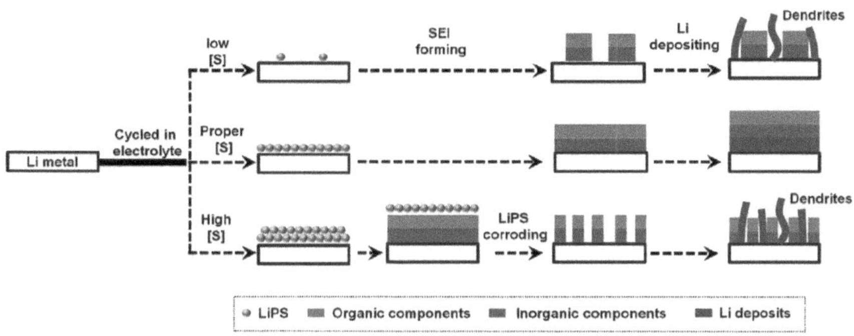

Fig. 34 Schematic describing the role of polysulfide concentration on SEI evolution and Li deposition [32].

the ether-based electrolyte could etch the SEI layer and lithium anode easily, resulting in a large consumption of organic electrolyte. The detailed process of this is shown in Fig. 34. Further enhancing of the areal mass loading and volumetric energy density of Li-S batteries can be obtained by increasing the mass loading and tap density of sulfur in the cathode, which increase the concentration of polysulfides in the electrolyte and areal current density on the anode surface.

To solve the corrosion problems of the anodes, protected anodes from metallic lithium were developed.

According to [33], several hosts for a Li metal anode for a Li-S cell demonstrated usefulness in suppressing Li dendrite growth, stabilizing the Li metal/electrolyte interface, reducing volume change, and improving the electrode efficiency; these include carbon nanotubes (CNTs), graphene, graphene-CNT hybrid, carbon nanofiber, porous carbon, graphite particles, graphite microtubes, TiC-carbon hybrid, 3D Cu current, Cu-Ni hybrid, metal foam, AlF_3 framework, Li alloys, and glass fibers. Such matrixes can regulate the nucleation and growth of the Li ion by lithiophilic sites and suppress Li dendrite growth by increased surface areas. The previously mentioned hosts are merely matrixes without Li in them and most of the advancements are mainly accomplished in Li metal anode half cells or full cells with the routine lithium-containing oxide or phosphate counter electrodes.

In [34], a 3D fibrous Li_7B_6 matrix as an inert host for elemental Li was presented, which was derived from the high-temperature alloying reaction of Li and B. Even at $10.0\,mA\,cm^{-2}$, no Li dendrite is observed, enabling a 2000-cycle lifespan in Li-S cells with a coulombic efficiency higher than 90%.

Another composite Li metal anode was obtained by a melting strategy [35–37]. First, a lithiophilic layer was coated onto the matrix to reduce the interfacial tension between the Li metal and the matrix. Then the matrix was immersed into molten Li and the molten Li was adsorbed into the matrix to obtain the composite Li anode [38].

Using LiF surface passivation on a 3D layered Li-rGO electrode, a composite Li metal anode was achieved, rendering Li-S cells with high safety, cyclability, coulombic efficiency, and rate capability (~800 mAh g^{-1} at 2 C) [39].

As mentioned in [33], the method for filling molten Li into the pores of host materials is complicated. The current material processing is conducted at a high temperature with a very sensitive regulation. The filling process is easily conducted when the lithiophilic hosts are applied, while the lithiophobic hosts must be further modified with a lithiophilic barrier.

8.2.1 Protected Anodes From Metallic Lithium

A stable, thin protective layer is preformed on the lithium surface by plasma polymerization, which helps to prevent the corrosion of the anode [40]. At room temperature, the protective layer has a low ionic conductivity of 3×10^{-8} S cm^{-1}, but this can be improved using polymerization by UV curing [41]. The protective layer reduces the parasitic chemical reaction of the polysulfide with the lithium anode, thus improving the cyclic capacity of the Li-S batteries. The Li_2CO_3 protective layer is induced on the surface by exposing the carbon dioxide electrode, which improves the charging-discharging behavior [42, 43]. The reaction between lithium and 1,3-dioxolane (DOL) forms a protective layer on the lithium surface, which extends the life of the lithium anode [44]. Alternatively, the LiAl alloy layer on the lithium surface effectively improves the life of a given Li-S battery [45], and recent discoveries indicate that $LiNO_3$ and organic nitrate additives in liquid electrolytes also prolong their lifetime [46]. These additives effectively modify SEI on the lithium electrode [47]. The transfer phenomenon is significantly reduced, thus increasing the use of sulfur and reducing self-discharge.

Recently, it was reported that thicker and less severe morphology (shown in Fig. 35C) of lithium fibers was deposited in the $LiNO_3$ electrolyte as compared to no $LiNO_3$, although $LiNO_3$ alone is not sufficient to prevent the growth of dendrites [31]. For comparison, a lower density of the emerging lithium particles was produced when Li_2S_8 and $LiNO_3$ were added together to the electrolyte, as shown in Fig. 35. No long fibers or dendrites could be found, even with a high deposition efficiency of 6 mAh cm^{-2} (Fig. 25F).

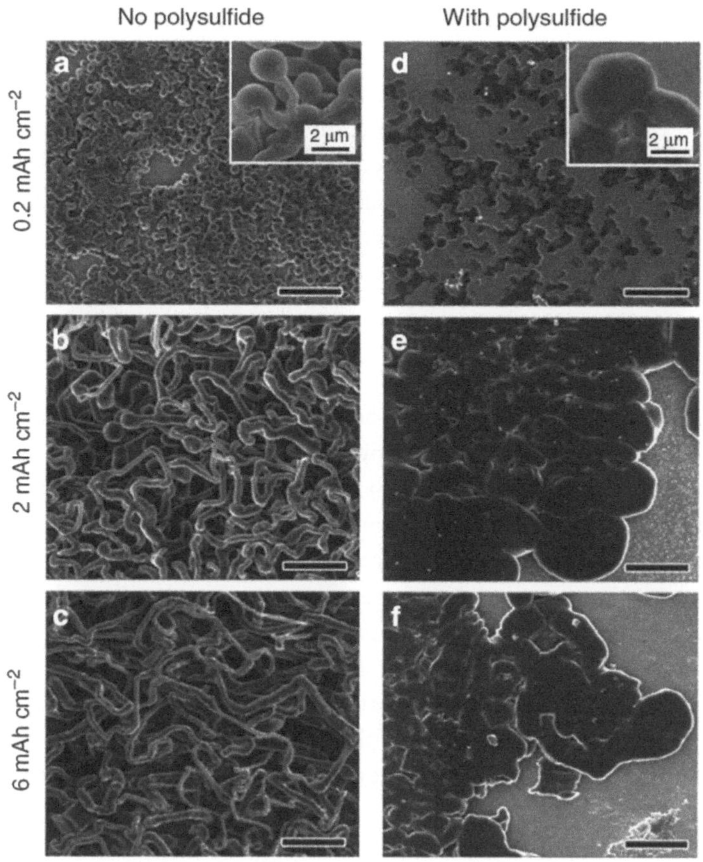

Fig. 35 Images from scanning electron microscope for metallic lithium deposited on pure stainless steel for an electrolyte with the addition of only $LiNO_3$ (A–C) and with the addition of $LiNO_3$ and Li_2S_8 (D–F) at different sediment volumes [31].

According to [33], surface coating is a facile and cost-effective method to deposit a protective layer onto a Li metal anode. Due to its simplicity, various coating methods were studied and nearly all the coating materials can suppress Li dendrite growth under certain conditions. Such coating materials included Al_2O_3, carbon, some polymers, and some alloys. The Li-S battery with a thin Li-Al alloy coating layer on the Li anode exhibits a better cycling performance than that with a pure Li anode.

As reported in [48], two-dimensional molybdenum disulfide (MoS_2) can also be used as a protective layer for Li-metal anodes to greatly improve the performance of Li-S batteries. Such a use of MoS_2 results in stable lithium electrodeposition and the suppression of dendrite nucleation sites.

The deposition and dissolution process of a symmetric MoS_2-coated Li-metal cell operates at a current density of $10\,mA\,cm^{-2}$ with low voltage hysteresis and a threefold improvement in cycle life, compared with using bare Li metal. For a Li-S full-cell configuration, using the MoS_2-coated Li as an anode and a 3D carbon nanotube–sulfur cathode, a specific energy of 589 Wh kg^{-1} and a coulombic efficiency of 98% were achieved for more than 1200 cycles at 0.5°C. As reported, molybdenum, when combined with two atoms of sulfur, creates a material that allows adjustment of the coating thickness, improving stability and compensating for the poor conductivity of sulfur, thus allowing for greater power density.

8.3 GRAPHITE ANODE

Deterioration of the lithium on the anode side due to formation of dendrites should be mitigated by appropriate anode protection or lithium alloying. Several attempts have been made to use carbons for this purpose. Particularly, a graphite anode has recently been developed as one of the alternative anode materials in Li-S batteries.

In [49], studies on graphite anodes were presented with the sulfur cathode being prelithiated electrochemically. After 60 cycles, 70% of the initial capacity was maintained. Although there was damage in the form of the weight proportion of graphite, a material with relatively low capacity, the use of graphite as the anode enabled general understanding of several well-developed lithium-ion batteries, which have had many successes in achieving excellent cyclical ability.

According to [49], a graphite anode, as much safer than a lithium metal anode, was used to fabricate Li/S cells with a nonlithiated sulfur composite cathode by incorporation of a lithium metal foil. The lithium foil provided lithium for cell performance. During cycling, the cells showed different voltage from that of a cell with a lithium metal anode.

Reference [50] presented a cell utilizing an electrolyte based on 1,2-dimethoxyethane (DME) solvent enabling reversible intercalation oflithium ions in graphite. A full cell consisting of a lithiated graphitic anode and lithium polysulfide cathode was constructed, which exhibited an initial capacity of 1500 mAh g^{-1} (based on the sulfur in the cathode) and a reversible capacity of 700 mAh g^{-1} for 100 cycles. This full cell could deliver over 460 mAh g^{-1} at rates as high as 2C. The cell degradation over prolonged cycles occurred due to the polysulfide shuttle, resulting in instability of the SEI layer on the graphitic anode.

The lithium anode can be also replaced by a composite anode containing graphite. One method to obtain the composite anode for Li-S cells is to electrochemically predeposit Li into the matrix. Reference [51] presented a covalently connected carbon nanostructure with CNTs growing on ultrathin graphite foam (CNT-UGF), which served as the current collector for both the sulfur cathode and lithium anode. The Li-S full battery adopting a sulfur/CNT-UGF cathode (47 wt% sulfur content in the whole cathode and 2.6 mg cm^{-2} sulfur loading) and lithium/CNT-UGF anode (20 wt% lithium content in the whole anode) exhibited a high capacity of 860 mAh g^{-1} at 12C (52 mA cm^{-2}) and a very low cycling capacity decay rate of 0.057% at 2.0C (8.7 mA cm^{-2}) after 400 cycles.

Contrary to the complicated routes for fabricating the composite Li metal anode, the introduction of an interlayer between the Li metal anode and membrane allows a simple regulation of the Li ion depositing behavior on the Li metal anode [52, 53].

In [54] a Li-S cell using electrically connected graphite and Li metal as a hybrid anode was presented. Lithiated graphite in front of the Li metal anode acted as an artificial, self-regulated SEI film. The functional SEI suppressed Li dendrite growth and minimized the deleterious side reactions between Li metal, LiPSs, and the electrolyte, rendering a significant performance improvement. Li-S batteries assembled with such a hybrid anode exhibited a discharge capacity of >800 mAh g^{-1} at a high current of 1737 mA g^{-1} with a coulombic efficiency >99% and only 11% capacity decay for 400 cycles.

Also, forms of carbon other than graphite were utilized for anodes in Li-S cells. For example, a carbon thin film with a continuous hemisphere (nanocaps) structure was used to cover the lithium surface [55]. Such a carbon interfacial layer was penetrated by Li^+ ions to promote the uniform nucleation of Li metal and subsequent Li granular growth within these nanocaps, rather than random Li nucleation forming sharp dendrites. As a result, such a covering allowed a metallic Li anode to charge and discharge at a high current density of 1 mAh cm^{-2} with a high coulombic efficiency over 99%.

Two-dimensional materials such as boron nitride or graphene sheets were also used, showing similar abilities in protecting Li metal against dendrite growth [56].

Another hybrid anode of Li metal and graphite was elaborated, where a graphite film was placed between the Li anode and the separator [54]. These two were then electrically connected to form a parallel anode and spontaneous Li insertion into the graphite interlayer occurred, providing a ready source of Li ions for further discharge. During cycling, the anode reaction occurred on the lithiated graphite interlayer rather than on the surface of the Li metal. This resulted in decreasing the irreversible reduction of polysulfides on Li metal in the Li-S cell and in better cell performance, including higher efficiency and stability.

The lithiated carbon can be utilized as an anode material instead of the metallic Li. Prelithiated hard carbon coated on carbon fibers was used as the anode in a Li-S battery during a study described in [57]. An initial discharge capacity of $1658\,mAh\,g^{-1}$ was achieved with a cathode composed of thermally infused sulfur in hollow carbon spheres, MWCNTs as the conductive additive, and PTFE as the binder. After 550 cycles, a capacity of $753\,mAh\,g^{-1}$ was retained, corresponding to 0.08% capacity fade per cycle. Although lithiated carbon can be used as an alternative anode in a Li-S cell, the low capacity of graphite/carbon materials in accommodating Li ions could possibly compromise the overall energy density of the cell.

8.4 SILICON ANODE

Silicon is a promising alternative to lithium as an anode for Li-S batteries, due to the high theoretical capacity of $\sim 3580\,mAh\,g^{-1}$ ($Li_{15}Si_4$). As is known, the practical application of Si-based anodes is limited by a serious loss of capacity, which is attributed to mechanical damage to the active material with a large volume change, >300%, during cyclic work [58–62]. This important problem has been solved using nanowire Si as the anode, which can control not only the change in volume, but also shorten the length of Li ion diffusion [63].

The intriguing high theoretical capacity of silicon attracts a lot of research attention to lithium-ion batteries [64, 65]. As stated in [16], the application of a silicon anode to Li-S batteries has been extended by using a prelithiated sulfur positive electrode.

Both the cathode (S) and the anode (Si) should be prelithiated during a full battery. Cui and his associates presented an anode from a silicon nanowire prelithiated by a simple self-discharge mechanism and constructed a full battery using a prelithiated silicon nanowire and sulfur as the anode and cathode, respectively.

Unfortunately, anodes based on silicon nanotubes do not relieve the rapid loss of Li-S battery capacity. Much greater success was achieved using a Sn-C nanocomposite as a negative electrode [66, 67]. Excellent cyclic performance was achieved at elevated temperatures using a polymeric electrolyte.

Fig. 36A shows the voltage profiles in the first, second, and tenth cycles for the sulfur cell/prelithiated silicon nanowire. Capacity retention of

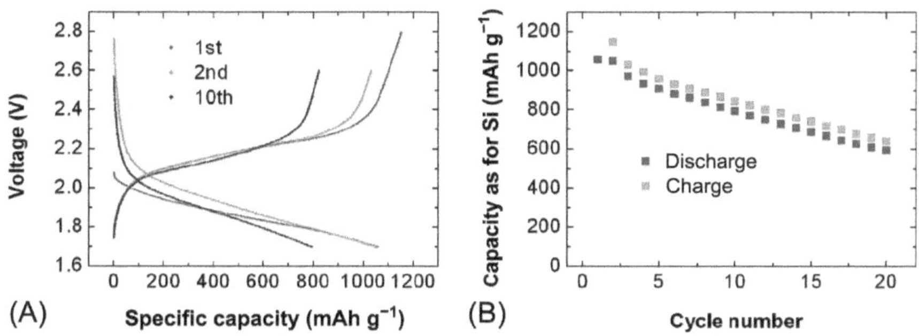

Fig. 36 (A) Voltage waveforms for the first, second, and tenth cycles of the full cell. (B) Cycling performance of the full cell [63].

around 80% could be achieved after 10 cycles (Fig. 36B), and severe capacity loss was observed during the first 20 cycles. However, it has been shown that a full battery can be made after using prelithiated anode materials and a sulfur-free pit cathode, which eliminates the safety problems associated with metallic lithium.

While currently mostly metallic lithium anodes are used in Li-S cells, the use of silicon anodes might offer better safety and durability [68]. However, in a lithium-sulfur-silicon (Li-S-Si) battery, lithium must be introduced either on the anode or on the cathode in the form of Li_2S. During the study, cells were used with Li_2S cathodes in combination with Si anodes (i.e., Si/Li_2S full cells) to investigate both the processes during initial charging/activation of Li_2S cathodes and the effect of Li_2S cathode activation on the cycling performance of Si/Li_2S full cells. The initial activation required a substantially higher charging potential than for the subsequent cycles. During the first cycle the gradual transformation of Li_2S to polysulfides and finally to crystalline sulfur was observed, i.e., even large Li_2S particles ($\approx$20 μm) could be charged completely. The formation of large sheets of sulfur at the cathode/separator interface was found. Similar cycling performance of Si/Li_2S full cells was observed at both 0.1 C and 1 C rates, with a clear advantage over Li/Li_2S cells, which suffered from severe dendrite formation at 1 C in the case of high Li_2S loadings.

REFERENCES

[1] M. Liu, X. Qin, Y.-B. He, B. Li, F. Kang, Recent innovative configurations in high-energy lithium–sulfur batteries, Mater. Chem. A 5 (2017) 5222–5234.

[2] MIT researchers discover two forms of lithium dendrite formation, Posted September 21, 2016 by Charles Morris & filed under Newswire, The Tech.

[3] M. Liu, D. Zhou, H. Jiang, Y. Ren, F. Kang, T. Zhao, A highly-safe lithium-ion sulfur polymer battery with SnO_2 anode and acrylate-based gel polymer electrolyte, Nano Energy 28 (2016) 97–105.
[4] X. Pu, G. Yang, C. Yu, Safe and reliable operation of sulfur batteries with lithiated silicon, Nano Energy 9 (2014) 318–324.
[5] X. Zhang, W. Wang, A. Wang, Y. Huang, K. Yuan, Z. Yu, J. Qiu, Y. Yang, Improved cycle stability and high security of Li-B alloy anode for lithium–sulfur battery, J. Mater. Chem. A 2 (2014) 11660–11665.
[6] M. Agostini, J. Hassoun, J. Liu, M. Jeong, H. Nara, T. Momma, T. Osaka, Y. Sun, B. Scrosati, A lithium-ion sulfur battery based on a carbon-coated lithium-sulfide cathode and an electrodeposited silicon-based anode, ACS Appl. Mater. Interfaces 6 (2014) 10924–10928.
[7] N. Moreno, M. Agostini, A. Caballero, J. Morales, J. Hassoun, A long-life lithium ion sulfur battery exploiting high performance electrodes, Chem. Commun. 51 (2015) 14540–14542.
[8] Z. Lu, N. Liu, H. Lee, J. Zhao, W. Li, Y. Li, Y. Cui, Nonfilling carbon coating of porous silicon micrometer-sized particles for high-performance lithium battery anodes, ACS Nano 9 (2015) 2540–2547.
[9] J. Wu, X. Qin, C. Miao, Y. He, G. Liang, D. Zhou, M. Liu, C. Han, B. Li, F. Kang, A honeycomb-cobweb inspired hierarchical core–shell structure design for electrospun silicon/carbon fibers as lithium-ion battery anodes, Carbon 98 (2016) 582–591.
[10] Q. Yun, X. Qin, Y. He, W. Lv, Y.V. Kaneti, B. Li, Q. Yang, F. Kang, Micron-sized spherical Si/C hybrids assembled via water/oil system for high-performance lithium ion battery, Electrochim. Acta 211 (2016) 982–988.
[11] J. Gao, M.A. Lowe, Y. Kiya, H.D. Abruña, Effects of liquid electrolytes on the charge–discharge performance of rechargeable lithium/sulfur batteries: electrochemical and in-situ X-ray absorption spectroscopic studies, J. Phys. Chem. C 115 (2011) 25132–25137.
[12] C. Xu, F. Lindgren, B. Philippe, M. Gorgoi, F. Björefors, K. Edström, Gustafsson, improved performance of the silicon anode for Li-ion batteries: understanding the surface modification mechanism of fluoroethylene carbonate as an effective electrolyte additive, Chem. Mater. 27 (2015) 2591–2599.
[13] Y. Yan, Y. Yin, S. Xin, J. Su, Y. Guo, L. Wan, High-safety lithium-sulfur battery with prelithiated Si/C anode and ionic liquid electrolyte, Electrochim. Acta 91 (2013) 58–61.
[14] J.R. Szczech, S. Jin, Nanostructured silicon for high capacity lithium battery anodes, Energy Environ. Sci. 4 (2011) 56–72.
[15] D. Zhou, M. Liu, Q. Yun, X. Wang, Y. He, B. Li, Q. Yang, Q. Cai, F. Kang, A novel lithiated silicon–sulfur battery exploiting an optimized solid-like electrolyte to enhance safety and cycle life, Small 13 (2017)1602015.
[16] Y. Yang, M.T. McDowell, A. Jackson, J.J. Cha, S.S. Hong, Y. Cui, New nanostructured Li_2S/silicon rechargeable battery with high specific energy, Nano Lett. 10 (2010) 1486–1491.
[17] S. Lee, S. Oh, E. Park, B. Scrosati, J. Hassoun, M. Park, Y. Kim, H. Kim, I. Belharouak, Y. Sun, Highly cyclable lithium–sulfur batteries with a dual-type sulfur cathode and a lithiated Si/SiO_x nanosphere anode, Nano Lett. 15 (2015) 2863–2868.
[18] P. Bai, J. Li, F.R. Brushett, M.Z. Bazant, Transition of lithium growth mechanisms in liquid electrolytes, Energy Environ. Sci. 9 (10) (2016) 3221–3229.
[19] A. Rosenman, R. Elazari, G. Salitra, E. Markevich, D. Aurbach, A. Garsuch, The effect of interactions and reduction products of $LiNO_3$, the anti-shuttle agent, in Li-S battery systems, J. Electrochem. Soc. 162 (2015) A470–A473.

[20] A. Jozwiuk, B.B. Berkes, T. Weiß, H. Sommer, J. Janek, T. Brezesinski, The critical role of lithium nitrate in the gas evolution of lithium–sulfur batteries, Energy Environ. Sci. 9 (2016) 2603–2608.
[21] S.S. Zhang, Role of $LiNO_3$ in rechargeable lithium/sulfur battery, Electrochim. Acta 70 (2012) 344–348.
[22] T. Zhang, K. Liao, P. He, H. Zhou, A self-defense redox mediator for efficient lithium–O_2 batteries, Energy Environ. Sci. 9 (2016) 1024–1030.
[23] G. Ma, Z. Wen, Q. Wang, C. Shen, J. Jin, X. Wu, Enhanced cycle performance of a Li–S battery based on a protected lithium anode, J. Mater. Chem. A 2 (2014) 19355–19359.
[24] G. Ma, Z. Wen, M. Wu, C. Shen, Q. Wang, J. Jin, X. Wu, A lithium anode protection guided highly-stable lithium–sulfur battery, Chem. Commun. 50 (2014) 14209–14212.
[25] L. Li, S. Basu, Y. Wang, Z. Chen, P. Hundekar, B. Wang, J. Shi, Y. Shi, S. Narayanan, N. Koratkar, Self-heating-induced healing of lithium dendrites, Science 359 (6383) (2018) 1513–1516.
[26] X.-W. Zhang, Y. Li, S.A. Khan, P.S. Fedkiw, Inhibition of lithium dendrites by fumed silica-based composite electrolytes, J. Electrochem. Soc. 151 (8) (2004) A1257–A1263.
[27] T. Tatsuma, M. Taguchi, N. Oyama, Inhibition effect of covalently cross-linked gel electrolytes on lithium dendrite formation, Electrochim. Acta 46 (8) (2001) 1201–1205.
[28] J. Steiger, G. Richter, M. Wenk, D. Kramer, R. Mönig, Comparison of the growth of lithium filaments and dendrites under different conditions, Electrochem. Commun. 50 (2015) 11–14. 188.
[29] R. Bouchet, Batteries: a stable lithium metal interface, Nat. Nanotechnol. 9 (8) (2014) 572–573.
[30] S.E. Cheon, K.S. Ko, J.H. Cho, S.W. Kim, E.Y. Chin, H.T. Kim, Rechargeable lithium sulfur battery. II. Rate capability and cycle characteristics, J. Electrochem. Soc. 150 (2003) A800–A805.
[31] W. Li, H. Yao, K. Yan, G. Zheng, Z. Liang, Y. Chiang, Y. Cui, The synergetic effect of lithium polysulfide and lithium nitrate to prevent lithium dendrite growth, Nat. Commun. 6 (2015) 7436.
[32] C. Yan, X. Cheng, C. Zhao, J. Huang, S. Yang, Q. Zhang, Lithium metal protection through in-situ formed solid electrolyte interphase in lithium-sulfur batteries: the role of polysulfides on lithium anode, J. Power Sources 327 (2016) 212–220.
[33] X.-B. Cheng, J.-Q. Huang, Q. Zhang, Review—Li metal anode in working lithium-sulfur batteries, J. Electrochem. Soc. 165 (1) (2018) A6058–A6072.
[34] X.-B. Cheng, H.-J. Peng, J.-Q. Huang, F. Wei, Q. Zhang, Small 10 (2014) 4257.
[35] Y. Liu, D. Lin, Z. Liang, J. Zhao, K. Yan, Y. Cui, Nat. Commun. 7 (2016) 10992.
[36] Z. Liang, D. Lin, J. Zhao, Z. Lu, Y. Liu, C. Liu, Y. Lu, H. Wang, K. Yan, X. Tao, Y. Cui, Proc. Natl. Acad. Sci. U. S. A. 113 (2016) 2862.
[37] D. Lin, Y. Liu, Z. Liang, H.-W. Lee, J. Sun, H. Wang, K. Yan, J. Xie, Y. Cui, Nat. Nanotechnol. 11 (2016) 626.
[38] C. Jin, O. Sheng, J. Luo, H. Yuan, C. Fang, W. Zhang, H. Huang, Y. Gan, Y. Xia, C.Liang, J. Zhang, X. Tao, Nano Energy 37 (2017) 177.
[39] D. Lin, Y. Liu, W. Chen, G. Zhou, K. Liu, B. Dunn, Y. Cui, Nano Lett. 17 (2017) 3731.
[40] Z. Takehara, Z. Ogumi, Y. Uchimoto, K. Yasuda, H. Yoshida, Modification of lithium/electrolyte interface by plasma polymerization of 1,1-difluoroethene, J. Power Sources 44 (1993) 377–383.
[41] Y.M. Lee, N.S. Choi, J.H. Park, J.K. Park, Electrochemical performance of lithium/sulfur batteries with protected Li anodes, J. Power Sources 119 (2003) 964–972.
[42] T. Osaka, T. Momma, Y. Matsumoto, Y. Uchida, Effect of carbon dioxide on lithium anode cycleability with various substrates, J. Power Sources 68 (1997) 497–500.

[43] T. Osaka, T. Momma, Y. Matsumoto, Y. Uchida, Surface characterization of electrodeposited lithium anode with enhanced cycleability obtained by CO_2 addition, J. Electrochem. Soc. 144 (1997) 1709–1713.
[44] Y. Nimon, S.J. Visco, M.-Y. Chu, Dioxolane as a proctector for lithium electrodes, 2001. US Patent 6, 225, 002.
[45] Y. Nimon, M.-Y. Chu, S.J. Visco, Coated lithium electrodes, 2003. US Patent 6, 537, 701.
[46] Y.V. Mikhaylik, Electrolytes for lithium–sulfur cells, 2008. US Patent 7, 354, 680.
[47] D. Aurbach, E. Pollak, R. Elazari, G. Salitra, C.S. Kelley, J. Affinito, On the surface chemical aspects of very high energy density, rechargeable Li–sulfur batteries, J. Electrochem. Soc. 156 (2009) A694–A702.
[48] E. Cha, M.D. Patel, J. Park, J. Hwang, V. Prasad, K. Cho, W. Choi, 2D MoS_2 as an efficient protective layer for lithium metal anodes in high-performance Li–S batteries, Nat. Nanotechnol. 13 (2018) 337–344.
[49] X. He, J. Ren, L. Wang, W. Pu, C. Wan, C. Jiang, Li/S lithium ion cell using graphite as anode, ECS Trans. 2 (2007) 47–49.
[50] A. Bhargav, M. Wu, Y. Fu, A graphite-polysulfide full cell with DME-based electrolyte, J. Electrochem. Soc. 163 (8) (2016) A1543–A1549.
[51] S. Jin, S. Xin, L. Wang, Z. Du, L. Cao, J. Chen, X. Kong, M. Gong, J. Lu, Y. Zhu, H.Ji, R.S. Ruoff, Covalently connected carbon nanostructures for current collectors in both the cathode and anode of Li–S batteries, Adv. Mater. 28 (2016) 9094–9102.
[52] Y. Liu, Q. Liu, L. Xin, Y. Liu, F. Yang, E.A. Stach, J. Xie, Making Li-metal electrodes rechargeable by controlling the dendrite growth direction, Nat. Energy 2 (2017)17083.
[53] Z. Wang, X. Wang, W. Sun, K. Sun, Dendrite-free lithium metal anodes in high performance lithium-sulfur batteries with bifunctional carbon nanofiber interlayers, Electrochim. Acta 252 (2017) 127.
[54] C. Huang, J. Xiao, Y. Shao, J. Zheng, W.D. Bennett, D. Lu, L.V. Saraf, M. Engelhard, L. Ji, J. Zhang, X. Li, G.L. Graff, J. Liu, Manipulating surface reactions in lithium–sulphur batteries using hybrid anode structures, Nat. Commun. 5 (2014) 3015.
[55] G. Zheng, S.W. Lee, Z. Liang, H.W. Lee, K. Yan, H. Yao, et al., Interconnected hollow carbon nanospheres for stable lithium metal anodes, Nat. Nanotechnol. 9 (2014) 618–623.
[56] K. Yan, H.W. Lee, T. Gao, G. Zheng, H. Yao, H. Wang, et al., Ultrathin two-dimensional atomic crystals as stable interfacial layer for improvement of lithium metal anode, Nano Lett. 14 (2014) 6016–6022.
[57] J. Brückner, S. Thieme, F. Böttger-Hiller, I. Bauer, H.T. Grossmann, P. Strubel, et al., Carbon-based anodes for lithium sulfur full cells with high cycle stability, Adv. Funct. Mater. 24 (2014) 1284–1289.
[58] N. Dimov, S. Kugino, M. Yoshio, Carbon-coated silicon as anode material for lithium ion batteries: advantages and limitations, Electrochim. Acta 48 (11) (2003) 1579–1587.
[59] H. Jung, M. Park, Y.-G. Yoon, G.-B. Kim, S.-K. Joo, Amorphous silicon anode for lithium-ion rechargeable batteries, J. Power Sources 115 (2) (2003) 346–351.
[60] X. Chen, K. Gerasopoulos, J. Guo, A. Brown, C. Wang, R. Ghodssi, J.N. Culver, Virus-enabled silicon anode for lithium-ion batteries, ACS Nano 4 (9) (2010) 5366–5372.
[61] X. Chen, K. Gerasopoulos, J. Guo, A. Brown, C. Wang, R. Ghodssi, J.N. Culver, A patterned 3D silicon anode fabricated by electrodeposition on a virus-structured current collector, Adv. Funct. Mater. 21 (2) (2011) 380–387.
[62] Y. Yao, K. Huo, L. Hu, N. Liu, J.J. Cha, M.T. McDowell, P.K. Chu, Y. Cui, Highly conductive, mechanically robust, and electrochemically inactive TiC/C nanofiber scaffold for high-performance silicon anode batteries, ACS Nano 5 (10) (2011) 8346–8351.
[63] N. Liu, L. Hu, M.T. McDowell, A. Jackson, Y. Cui, Prelithiated silicon nanowires as an anode for lithium ion batteries, ACS Nano 5 (8) (2011) 6487–6493.

[64] C.K. Chan, H.L. Peng, G. Liu, K. McIlwrath, X.F. Zhang, R.A. Huggins, Y. Cui, High-performance lithium battery anodes using silicon nanowires, Nat. Nanotechnol. 3 (2008) 31–35.
[65] L.F. Cui, R. Ruffo, C.K. Chan, H.L. Peng, Y. Cui, Crystalline-amorphous core – shell silicon nanowires for high capacity and high current battery electrodes, Nano Lett. 9 (2009) 491–495.
[66] J. Hassoun, B. Scrosati, Moving to a solid-state configuration: a valid approach to making lithium-sulfur batteries viable for practical applications, Adv. Mater. 22 (2010) 5198–5201.
[67] J. Hassoun, B. Scrosati, A high-performance polymer tin sulfur lithium ion battery, Angew. Chem. Int. Ed. 49 (2010) 2371–2374.
[68] H. Jha, I. Buchberger, X. Cui, S. Meini, H.A. Gasteiger, Li-S batteries with Li_2S cathodes and Si/C anodes, J. Electrochem. Soc. 162 (9) (2015) A1829–A1835.

CHAPTER NINE

Separators

In comparison with intensive research on the construction of electrode material, only a few scientists are working on separators for Li-S batteries, which are considered a key element in the battery system [1–3].

The main function of the separators in these batteries is to conduct the ions and block the electrons to perform reversible charging/discharging and to prevent direct contact between the electrodes: cathode and anode. It is expected that the ideal separator for Li-S batteries not only has good ionic conductivity, but also inhibits polysulfide migration during cyclic work [4–6]. As described in [7–9], the separator is used to prevent short circuits between the electrodes. It is tightly connected to the cathode and often saturated with organic electrolyte. According to [10], developing functional separators can solve both the shuttle effect and low mass loading concern for Li-S cells.

Yao et al. presented a conductive coating on the separator to prevent the accumulation of inactive, sulfur-derived molecules in the cathode [1]. The cyclic efficiencies of cells with a primary separator (polypropylene, PP) and coatings of different materials on the separator are shown in Fig. 37. The initial capacity of $1100\,mAh\,g^{-1}$ can be achieved for a Ketjenblack cell and a separator coated with multiwall carbon nanotubes. In contrast, only $\sim 700\,mAh\,g^{-1}$ was obtained for a cell with a Super P coated separator. This was probably due to the higher conductivity of Ketjenblack and multiwall carbon nanotubes in relation to Super P. It should be noted that a significant improvement was achieved for the separator cell covered with Super P as compared to the primary separator.

Although microporous polyolefin membranes are currently the most widely used separators for lithium batteries, their low porosity and poor wettability with liquid electrolyte increase the resistance of the cell, resulting in poor kinetics and low battery capacity. Poor cell capacity with primary polypropylene is shown in Fig. 38. Even for a cell with a Super P–covered separator, a capacity of $\sim 300\,mAh\,g^{-1}$ can be obtained only at a current density of 1C. This was probably due to the low porosity and poor wettability of the liquid electrolyte of the microporous polyolefin membrane, resulting in poor battery kinetics.

https://doi.org/10.1016/B978-0-12-816392-4.00009-8

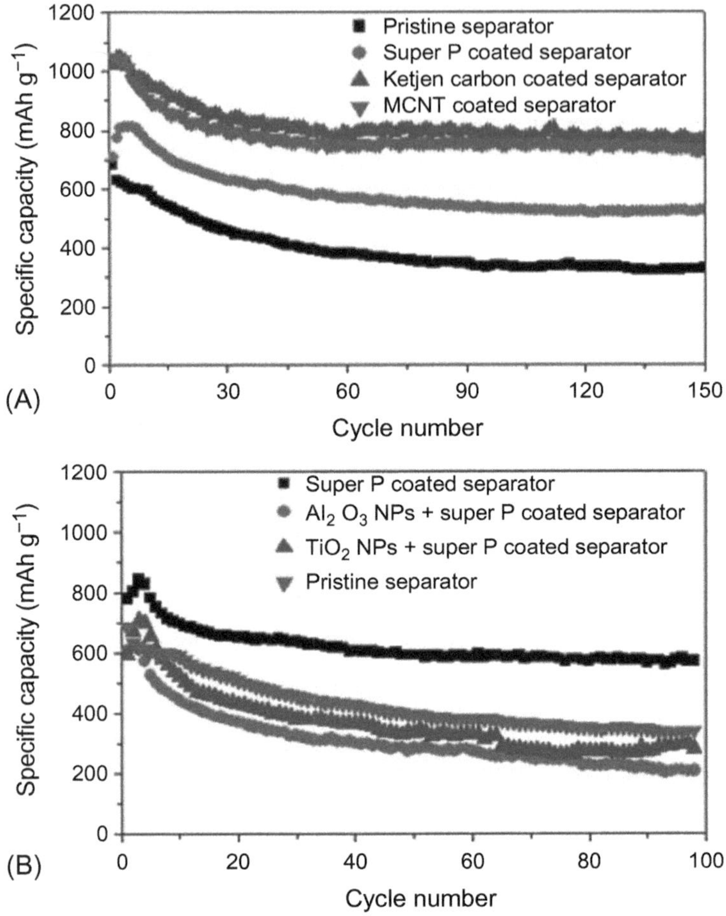

Fig. 37 Cyclic capacity of cells with the primary separator and coatings of various materials on the separator at C/5 for a sulfur content of 60% by weight. (A) Cycling stability comparison of Li-S cells with different carbon nanomaterial coatings. (B) Cycling stability comparison of Li-S cells with different metal oxide nanomaterial coatings [1].

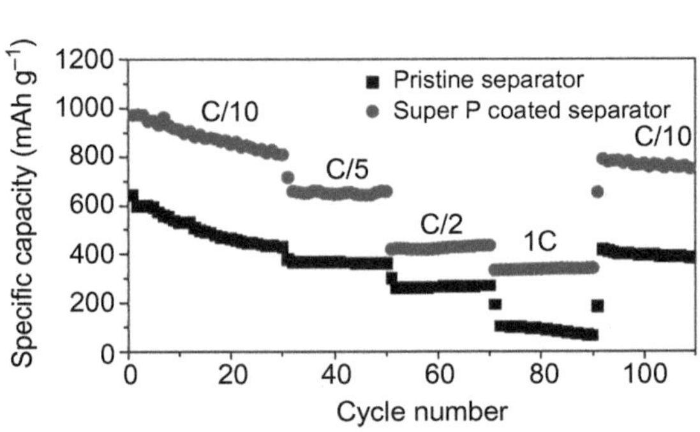

Fig. 38 Rate capability of Li-S cells with pristine separator (50 wt% cathode) and Super P modified separator (60 wt% cathode) [1].

Therefore, to achieve a better capacity for Li-S batteries, the proper separator should not only have a very porous structure and excellent electrolyte wettability, but should also be able to capture sulfur-derived particles during cyclic work, increasing the use of the active material.

Initially, as described in [11], a commercial separator was modified by coating a Super P carbon layer on one side. The coated carbon layer was able to function as a conductive upper current collector and a polysulfide-diffusion barrier region, whereas the commercial polymer separator served as an electrical insulating layer.

In [5] a Nafion layer on a PP/PE/PP separator was introduced to improve the stability and coulombic efficiency of Li-S batteries. The SO_3^- group–coated channels allow ion hopping of positively charged species (Li^+) but reject hopping of the negative ions of polysulfides due to the coulombic interactions, as shown in Fig. 39. Low ionic conductivity still encumbers the development of this technique, particularly when the areal sulfur loading reaches an important level.

According to [12], separators that can block the mossy growth are made of anodic alumina with a thickness of 60 μm and have well-fitting, simple nanopores throughout their thickness. This explains why sometimes better performance occurs with the use of ceramic separators. Research results indicate that elastic composite ceramic separators, such as those prepared by coating ceramic molecules with conventional polyolefin separators, should be used to block the root-growing mossy lithium.

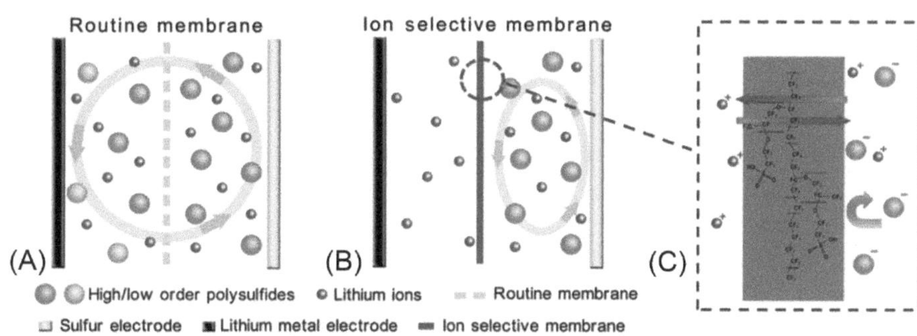

Fig. 39 Schematic of the different Li-S battery configurations. (A) Cell with routine membranes, in which polysulfides shuttle between the cathode and anode sides. (B) Cell with ion selective membranes, in which the polysulfide anions are limited to the cathode side. (C) Enlarged schematic showing that the cation perm-selective membrane allows the transport of lithium cations and blocks polysulfide anions. *(Taken from J.-Q. Huang, Q. Zhang, H.-J. Peng, X.-Y. Liu, W.-Z. Qian, F. Wei, Ionic shield for polysulfides towards highly-stable lithium–sulfur batteries, Energy Environ. Sci. 7(1) (2014) 347–353.)*

According to [10], coating the separator by a thin layer of functional materials (carbon, polymer, and oxide) on the cathode side or grafting specific functional groups on it could effectively restrain the shuttle effect of Li-S batteries, and consequently achieve satisfactory performances. The configuration of Li-S systems with a modified separator is shown in Fig. 40. The functionalization technique can enhance the ionic selectivity of the separator, analogously to the ion-selective membranes in fuel cells or redox flow batteries. Some materials, like Super P, graphene, mesoporous carbon, multiwalled carbon nanotubes (MWCNT)@PEG, Nafion, polydopamine, glassy fiber paper, black phosphorus, sulfonated acetylene black, and carboxyl functional groups, were all introduced to decorate commercial separators and improve the immobilization function.

According to [13], such ionic perm-selective layers and groups inhibit the free diffusion of polysulfides and simultaneously protect the lithium anode from parasitic reactions by sulfur species.

A graphene-protected separator was presented in [8] to construct a sandwich-structured sulfur cathode, in which the graphene-separator bilayer efficiently decreased the contact impedance of the entire device. The carbon layer could work as a physical barrier to block the diffusion of polysulfides.

According to [14], both the ionic conductivity and selectivity could be enhanced by grafting facile carboxyl functional groups on the commercial PP separator.

Therefore, the polysulfide-shuttle behavior in Li-S cells can be suppressed, and the cells can also deliver excellent rate capability because the pores of the separator, serving as channels for ion transport, are well maintained during the modification process.

According to [10] there is also a feasible method to fabricate multifunctional separators with high ionic conductivities and considerable polysulfide blocking ability by coating multicomponents on the separators.

As reported in [15], a dense hybrid coating was developed composed of orientation-aligned graphene layers and uniformly embedded $Li_4Ti_5O_{12}$ nanoparticles on the separator via a vacuum-filtration method. The coating layer achieved an ideal combination of physical barrier and chemical adsorption for polysulfide diffusion and simultaneously maintained fast lithium ion transport. The dense hybrid coating reached a balance of blocking function for polysulfides and fast lithium ion transport, showing a better perspective than the loose coating prepared by the blade-coating method, as shown in Fig. 41.

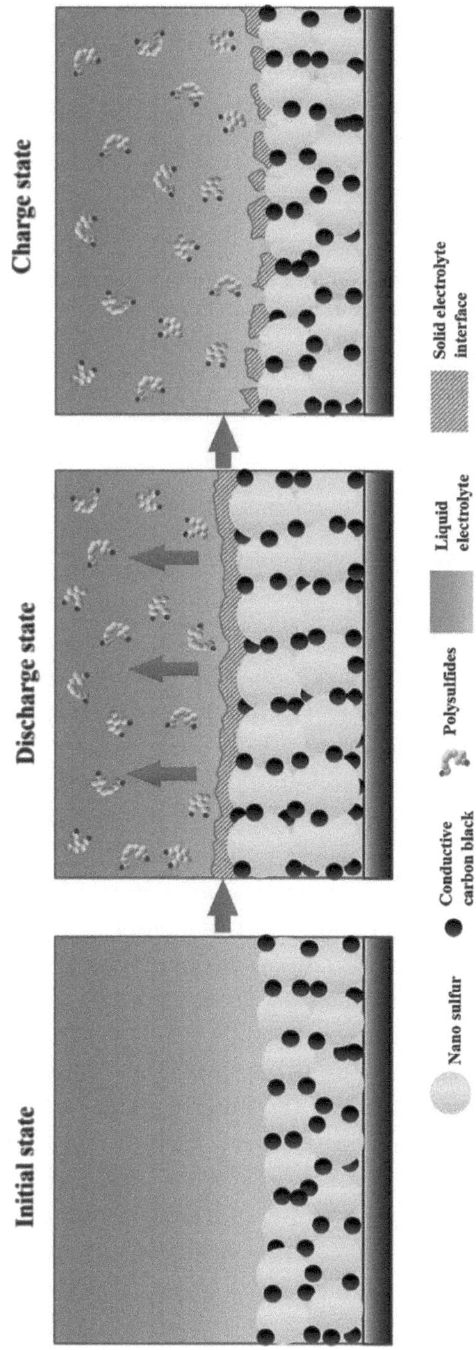

Fig. 40 Configuration of a Li-S battery with a modified separator to inhibit the detrimental diffusion of polysulfides to the anode side. *(Taken from M. Liu, X. Qin, Y.-B. He, B. Li, F. Kang, Recent innovative configurations in high-energy lithium–sulfur batteries, J. Mater. Chem. A 5 (2017) 5222.)*

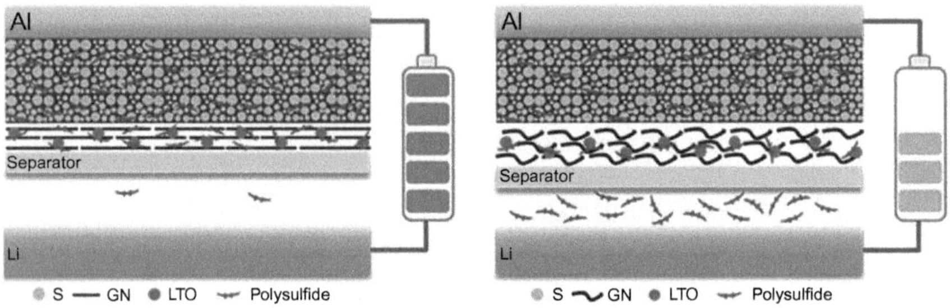

Fig. 41 Schematic of Li-S batteries with a G-LTO/PP separator produced by vacuum-filtration (left, dense coating layer) and blade coating (right, loose coating layer). *(Taken from M. Liu, X. Qin, Y.-B. He, B. Li, F. Kang, Recent innovative configurations in high-energy lithium–sulfur batteries, J. Mater. Chem. A 5 (2017) 5222.)*

During studies described in [16], a ternary-layered separator with a macroporous PP matrix layer, GO barrier, and Nafion retarding layer was elaborated as the separator for Li-S batteries by the vacuum-filtration method. In the ternary-layered separator, the ultrathin GO layer blocked the macropores of the PP matrix, and a dense ion selective Nafion layer was attached as a retarding layer to suppress the crossover of sulfur-containing species. The ternary-layered separators improved the initial capacity and coulombic efficiency of Li-S batteries.

As reported in [17], a sustainable and highly porous PAN/GO nanofiber membrane separator was fabricated based on an electrospinning technique, enabling high capacity and excellent anti-self-discharge capability for Li-S cells. The high performances resulted from the highly porous structure and excellent electrolyte wettability of the nanofiber separator, as well as the excellent barrier effects caused by the high energy binding between the cyano group and sulfur species and the electrostatic interaction between GO and negatively charged species.

In [18] a metal–organic framework (MOF)/GO composite film was presented as a separator to mitigate the shuttling problem for Li-S batteries. Due to the small MOF pore size (about 0.1 nm), the MOF/GO separator could act as an ionic sieve in Li-S cells, which selectively sieves Li^+ ions while efficiently suppressing undesired polysulfides migrating to the anode side. As a result, Li-S batteries with the MOF/GO separator exhibited sustainably enhanced cycling stability and a lower capacity-fading rate. This strategy can be useful for the development of novel separators with ion sieve functions for energy storage devices.

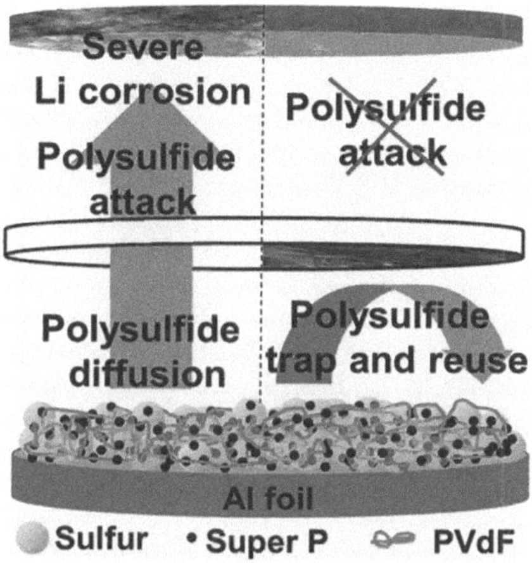

Fig. 42 The improvement mechanism of the SWCNT-modulated separator. *(Taken from C.-H. Chang, S.-H. Chung, A. Manthiram, Effective stabilization of a high-loading sulfur cathode and a lithium-metal anode in Li-S batteries utilizing SWCNT-modulated separators, Small 12(2) (2016) 174–179.)*

As reported in [19], a custom single-wall carbon nanotube (SWCNT)–modulated separator was employed to directly suppress the polysulfide migration and indirectly protect the lithium-metal anode from severe polysulfide contamination (Fig. 42). The conductive sp^2-carbon scaffold continuously reactivated and reutilized the trapped active material, so the SWCNT-modulated separator provided a simple way to facilitate the implementation of pure sulfur cathodes with high sulfur contents and loadings.

REFERENCES

[1] H. Yao, K. Yan, W. Li, G. Zheng, D. Kong, Z.W. Seh, V.K. Narasimhan, Z. Liang, Y. Cui, Improved lithium–sulfur batteries with a conductive coating on the separator to prevent the accumulation of inactive S-related species at the cathode–separator interface, Energy Environ. Sci. 7 (10) (2014) 3381–3390.

[2] S.-H. Chung, A. Manthiram, High-performance Li–S batteries with an ultra-lightweight MWCNT-coated separator, J. Phys. Chem. Lett. 5 (11) (2014) 1978–1983.

[3] J. Fang, F. Qin, J. Li, K. Zhang, W. Liu, M. Wang, F. Yu, L. Zhang, Improved performance of sulfur cathode by an easy and scale-up coating strategy, J. Power Sources 297 (2015) 265–270.

[4] W. Cai, G. Li, F. He, L. Jin, B. Liu, Z. Li, A novel laminated separator with multi functions for high-rate dischargeable lithium–sulfur batteries, J. Power Sources 283 (2015) 524–529.

[5] J.-Q. Huang, Q. Zhang, H.-J. Peng, X.-Y. Liu, W.-Z. Qian, F. Wei, Ionic shield for polysulfides towards highly-stable lithium–sulfur batteries, Energy Environ. Sci. 7 (1) (2014) 347–353.

[6] R. Song, R. Fang, L. Wen, Y. Shi, S. Wang, F. Li, A trilayer separator with dual function for high performance lithium–sulfur batteries, J. Power Sources 301 (2016) 179–186.
[7] J. Hassoun, J. Kim, D. Lee, H. Jung, S. Lee, Y. Sun, B. Scrosati, A contribution to the progress of high energy batteries: a metal-free, lithium-ion, silicon–sulfur battery, J. Power Sources 202 (2012) 308–313.
[8] G. Zhou, S. Pei, L. Li, D. Wang, S. Wang, K. Huang, L. Yin, F. Li, H. Cheng, A graphene–pure-sulfur sandwich structure for ultrafast, long-life lithium–sulfur batteries, Adv. Mater. 26 (2014) 625–631.
[9] Y. Yin, S. Xin, Y. Guo, L. Wan, Lithium-sulfur batteries: electrochemistry, materials, and prospects, Angew. Chem. Int. Ed. 52 (2013) 13186–13200.
[10] M. Liu, X. Qin, Y.-B. He, B. Li, F. Kang, Recent innovative configurations in high-energy lithium–sulfur batteries, J. Mater. Chem. A (5) (2017) 5222.
[11] S. Chung, A. Manthiram, Bifunctional separator with a light-weight carbon-coating for dynamically and statically stable lithium-sulfur batteries, Adv. Funct. Mater. 24 (2014) 5299–5306.
[12] Charles Morris. MIT Researchers Discover Two Forms of Lithium Dendrite Formation, Posted September 21, 2016 by filed under Newswire, The Tech.
[13] J. Balach, T. Jaumann, M. Klose, S. Oswald, J. Eckert, L. Giebeler, Functional mesoporous carbon-coated separator for long-life, high-energy lithium–sulfur batteries, Adv. Funct. Mater. 25 (2015) 5285–5291.
[14] X. Yu, J. Joseph, A. Manthiram, Suppression of the polysulfide-shuttle behavior in Li–S batteries through the development of a facile functional group on the polypropylene separator, Mater. Horiz. 3 (2016) 314–319.
[15] Y. Zhao, M. Liu, W. Lv, Y. He, C. Wang, Q. Yun, B. Li, F. Kang, Q. Yang, Dense coating of Li4Ti5O12 and graphene mixture on the separator to produce long cycle life of lithium-sulfur battery, Nano Energy 30 (2016) 1–8.
[16] T. Zhuang, J. Huang, H. Peng, L. He, X. Cheng, C. Che, Q. Zhang, Rational integration of polypropylene/graphene oxide/nafion as ternary-layered separator to retard the shuttle of polysulfides for lithium–sulfur batteries, Small 12 (2016) 381–389.
[17] J. Zhu, C. Chen, Y. Lu, J. Zang, M. Jiang, D. Kim, X. Zhang, Highly porous polyacrylonitrile/graphene oxide membrane separator exhibiting excellent anti-self-discharge feature for high-performance lithium–sulfur batteries, Carbon 101 (2016) 272–280.
[18] S. Bai, X. Liu, K. Zhu, S. Wu, H. Zhou, Metal–organic framework-based separator for lithium–sulfur batteries, Nat. Energy 1 (2016) 16094.
[19] C.-H. Chang, S.-H. Chung, A. Manthiram, Effective stabilization of a high-loading sulfur cathode and a lithium-metal anode in Li-S batteries utilizing SWCNT-modulated separators, Small 12 (2) (2016) 174–179.

CHAPTER TEN

Sulfur Pouch Cells

According to [1], in a sulfur pouch cell conductive foil-tabs are welded to the electrodes and brought to the outside in a fully sealed way. The pouch cell can be a simple, flexible, and lightweight solution to battery design. Some stack pressure is needed but allowance for swelling must be made. The pouch cells can deliver high load currents, but they achieve the highest performance under light loading conditions and with moderate charging.

The pouch cell achieves 90%–95% packaging efficiency, the highest among battery packs. Eliminating the metal enclosure reduces weight, but the cell needs support and allowance to expand in the battery compartment. There are many design solutions for pouch cells, and their packs are used in consumer, military, and automotive applications.

Pouch packs are commonly Li-polymer. Small cells are used for portable applications requiring high load currents, such as drones. The larger cells, in the 40 Ah range, serve in energy storage systems (ESS), because fewer cells simplify the battery design.

Although easily stackable, swelling must be prevented. While smaller pouch packs can grow 8%–10% over 500 cycles, large cells may expand to that size in 5000 cycles. It is best not to stack pouch cells on top of each other but to lay them flat, side by side, or allow extra space between them. Sharp edges should be avoided, as they can stress the pouch cells as they expand.

Up to 3% swelling incidents were reported for pouch packs on a poor batch run. The pressure created can crack the battery cover or even break the display and electronic circuit boards. Using an inflated battery and puncturing the bloating cell near heat or fire is prohibited, as the escaping gases can ignite. Swelling can occur due to gassing but is less dangerous for large pouch cell designs. The gases contain mainly CO_2 (carbon dioxide) and CO (carbon monoxide).

Pouch cells are manufactured by adding a temporary "gasbag" on the side. Gases escape into the gasbag while forming the solid electrolyte interface (SEI) during the first charge. The gasbag is cut off and the pack is resealed as part of the finishing process. The gas generation, called gassing,

Next-generation Batteries with Sulfur Cathodes
https://doi.org/10.1016/B978-0-12-816392-4.00010-4

cannot be fully avoided. It is caused by electrolyte decomposition as part of usage and aging. Stresses, such as overcharging and overheating, promote gassing. Ballooning of pouch cells with normal use can be a hint of a flawed batch.

The pouch cells have the potential for greater capacity than the cylindrical format. Large flat packs serve electric powertrains and ESSs with good results. The cost per kilowatt-hour for the pouch cell is still higher than with the cylindrical cell, but this is changing.

According to [2], so far, at the pouch cell level, Li-S batteries can achieve a practical gravimetric energy density E_g of 350 Wh kg^{-1} with a cycling life of 100 cycles.

According to [2], the Sion Power manufacturer announced that their Li-S pouch batteries, with an E_g of 350 Wh kg^{-1}, successfully assisted the 11-day flight of a solar/battery powered aircraft (Zephyr 7) in winter. But no details on the cycle life and volumetric energy density were provided. The Oxis Energy manufacturer developed a Li-S pouch battery with an E_g of 300 Wh kg^{-1}, which could successfully be cycled for 80 cycles; however, its volumetric energy density E_v was below 200 Wh L^{-1}, much lower than that of commercial lithium-ion batteries (LIBs).

In [2] the prototype of a pouch-type Li-S cell using sulfur-carbon powder (88% sulfur content) was also presented. For the slurry preparation, a water-based binder (LA132) and Super-P conductive carbon were used. Typically, the thickness of the double-side–coated cathode was 250–300 μm. The sulfur loading ranged from 4.5 to 6.5 mg cm^{-2} with 75% sulfur content. According to the model, the calculated porosity of the electrode was 71%–76%. The electrolyte/sulfur ratio was 3 μL mg^{-1} in the case. The thickness of the lithium foil used was 150–200 μm corresponding to 200% excess. No Cu current collector was used, and the lithium foil worked as a double-sided anode. Its E_v was ~300 Wh L^{-1}. It is noted that some electrolyte was pumped out from the stacked electrode to the empty part of the package during discharge and pumped back into the electrode in the following charging process. The movement of the electrolyte was caused by the volume change of the sulfur electrode during the discharge/charge processes. That phenomenon disappeared after several cycles, possibly because of the high porosity mossy lithium after cycling, which stored the electrolyte pumped out from the cathode.

In Ref. [3], many aspects of the manufacturing of a 3 Ah Li-S pouch cell were discussed. It was found that when coin cell tests are adapted to the pouch cell, many pronounced problems are detected, resulting in much faster cell failure. Initially the pouch cell exhibited 120 Wh kg^{-1} energy density based on the mass of the entire cell and loses its almost 85% capacity after 100 cycles. The relatively low gravimetric energy density and the short cycle life of the Li-S cells is directly linked to lower sulfur utilization, the unprotected Li metal anode, and the need to operate with more electrolyte than 60 wt% of the entire cell, resulting in the decrease of energy density of the investigated Li-S system.

During electrochemical tests described in [4], the Li-S pouch cell (Fig. 43) was assembled using the solution of 0.1 M lithium trifluoromethanesulfonate ($LiCF_3SO_3$) and 0.1 M lithium nitrate ($LiNO_3$) dissolved in dimethoxyethane/1,3-dioxolane (DME/DOL, 1:1 by volume) as the electrolyte. A pure sulfur cathode (area: 12 cm^2) with an aluminum tab was placed with an upper carbon current collector over it. A separator (Celgard 2400) was placed over the carbon upper current collector and a lithium foil (with a nickel tab) was placed on top of it. The pouch cell was sealed with an aluminum soft packaging film. The electrolyte-to-sulfur ratio for the pouch cell was 8. The galvanostatic discharge/charge tests were carried out with an Arbin battery test station at different C rates ($1C = 1675$ mA g^{-1}), and all the capacity values were calculated based on the weight of sulfur in the cathode. Galvanostatic intermittent titration

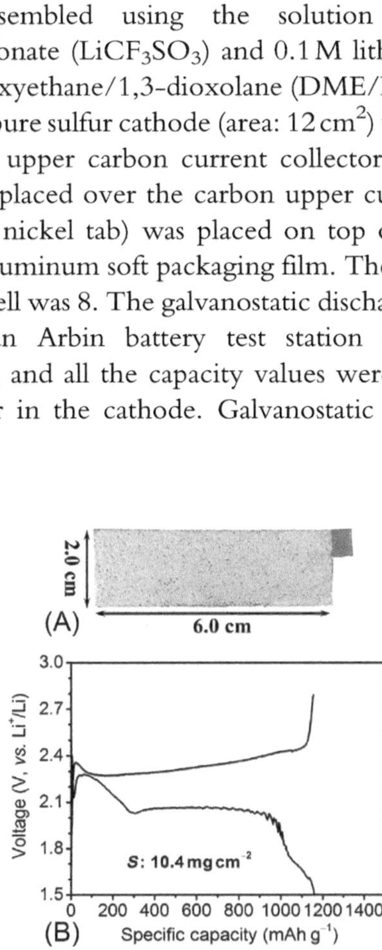

Fig. 43 (A) Photograph of a pure sulfur electrode in pouch cell configuration; (B) initial discharge/charge curves of a pouch cell with pure sulfur cathode cycled at a C/20 rate. *(Taken from L. Qie, A. Manthiram, High-energy-density lithium–sulfur batteries based on blade-cast pure sulfur electrodes, ACS Energy Lett. 1(1) (2016) 46–51, DOI: 10.1021/acsenergylett.6b00033.)*

technique (GITT) profiles were recorded by applying a constant current (C/50) for 1 h, followed by a resting step of 1 h.

In [5] an approach allowing a state of charge determination of lithium-sulfur batteries using in-situ impedance spectroscopy was presented. The impedance of lithium-sulfur pouch cells was measured at different states of charge and temperature to determine the influences of these states on the battery impedance. It was shown that impedance measurements can be used for a more accurate state of charge determination than conventional methods, like open circuit voltage detection. Also, an algorithm for state of charge determination was developed and implemented into an impedance measurement–ready battery management system.

According to [6], a novel electrolyte approach for a lithium-sulfur secondary battery was elaborated, realistically portrayed on a pouch cell level, with its decisive element being a new electrolyte system, a mixture of a hydrofluoroether and sulfolane. The hereby suppressed polysulfide solubility enables high discharge capacities and requires only very low electrolyte excess ($<2.6\,\mu L\,mg S^{-1}$), achieving high coulombic efficiencies above 94% (capacity retention of 77% over 40 cycles) without the necessity of adding lithium nitrate for shuttle suppression. Due to the low volatility, this adopted electrolyte concept promises significant benefits for large operational battery units.

During studies described in [7] pouch cells were used, assembled (in dry room) by stacking the lithium foil (ChemetallFoote Corp., 50 mm), separator, and cathode. The electrolyte was a solution of 8 wt% lithium bis (trifluoromethanesulfonyl)imide (Aldrich, 99.95%), 4 wt% lithium nitrate (Merck, 99.995%), 44 wt% 1,2-dimethoxyethane (Alfa Aesar, 99+%) and 44 wt% 1,3-dioxolane (Acros, 99.8%). Prior to use, the solid electrolyte components were dried in vacuum and the solvents were purified by distillation from sodium potassium alloy under argon. The water content in the electrolyte was as low as 33 ppm, as determined by Karl-Fischer titration. The electrolyte-to-sulfur mass ratio was 10:1 for the pouch cells used.

It was found that simple cathode design (cathodes with a nonoptimized composite microstructure) provided good electrochemical performance both in coin and pouch cells with sulfur loadings of $2\,mg\,cm^{-2}$. It was demonstrated that:

(1) specific capacities of $1000\,mA\,h\,g^{-1}$ can be achieved over 60 cycles at room temperature while the cyclability at elevated temperatures (here, $\vartheta > 40°C$) is poor,

(2) the discharge is the kinetically rate-limiting process,

(3) the major fraction of active sulfur in the electrode is lost during the formation cycle at C/50, and

(4) the Li-S cells suffer from drying-out due to continuous electrolyte decomposition on the lithium metal anode.

The Li_2S formation (grain size of 10 nm) on discharge and the appearance of single phase b-sulfur in the sub-100 nm size range were also found—rather than the thermodynamically stable orthorhombic polymorph (a-sulfur)—by the end of the charge cycle.

According to [8], existing prototype cells are still far from the target (~600 Wh kg^{-1}) but cell manufacturers are expecting to hit it soon.

The OXIS Energy prototype [9] of a rechargeable Li-S pouch cell had a Li metal anode. The cell with dimensions 176 mm × 114 mm × 13.7 mm was characterized by the maximum voltage of 2.45 V, a nominal voltage of 2.05 V, and capacity of 21 Ah (at 0.2 C, 30°C). Its weight was 233 ± 5 g and volume was 0.252 L. The recommended conditions for the cell were the following:

- for charging: 2.1 A constant current (C/10) at 30°C. This corresponds to the situation when charge stops at 2.45 V or 11 h max charge time occurred,
- for discharging: a peak discharge current of 2 C, a maximum continuous discharge current (at 30°C) of 1.5 C.

The cell maximum continuous power was 60 W at its optimum operating temperature of 30°C. The cycle number before arriving at 80% of the beginning-of-life capacity was 180 cycles, when cycled at 30°C in a standard life test regime.

The other prototype Li-S pouch cell, described in [10], was composed of a sulfur positive electrode cast on NwC, lithium foil as the negative electrode, two layers of porous separator (Celgard® 2400 and Viledon®) soaked in liquid organic electrolyte: 1 mol L^{-1} lithium bis(trifluoromethane) sulfonimide (LiTFSI, Rhodia) in a 1:1 volume mixture of tetraethylene glycol dimethylether (TEGDME, 99%, Aldrich) and 1,3-dioxolane (DIOX, anhydrous, 99.8%, Aldrich), with addition of 0.1 mol L^{-1} lithium nitrate ($LiNO^3$, 99.99%, Aldrich). To eliminate the signal from aluminum-laminated pouches, three holes were cut (Ø 3 mm) and Kapton® tape was stuck from both sides. To avoid any air penetration through the Kapton® layer, the cell was tightly sealed in a polyethylene bag. Each electrode had a hole (Ø 3 mm), so that we could have the response from both positive and negative electrodes separately. Viledon® was used as an additional layer of separator, which served as reservoir for a sufficient amount of electrolyte in the cell. It had holes in exact positions as packaging foil, not to shield the eventual peaks of crystalline Li_2S and sulfur appearing during cycling. To maintain good electrical contact inside the prototype, a moderate pressure was applied by two parallel plastic plates (with three holes, Ø 3 mm) pinned to the cell with metal clips. The pouch cells were assembled in a dry room (−40°C dew point) and activated with electrolyte in an argon-filled glove box (H_2O, $O_2 < 0.1$ ppm).

A very important issue is the modeling of pouch cells. During studies described in [11] the equivalent circuit model (ECM) for pouch cells, based on the equivalent circuit proposed in [12] for a Li-S cell in an intermediate state, was used. For the ECMs parametrized based on EIS measurements, constant phase elements (CPEs) are usually used instead of capacitors to consider a nonideal behavior of the electrode, like a roughness of the surface and porosity of a material [13]. If the parametrization is performed based on current pulse measurements, then the utilized model contains only capacitors instead of CPEs. The layout of the ECM used in this work for modeling the dynamic behavior of the Li-S cell is presented in Fig. 44. According to [12], R_0 represents the electrolyte resistance, R_1 stands for the total surface layer resistance of the sulfur and lithium electrodes, C_1 is the distributed surface layer capacitance on both electrodes, R_2 expresses the charge transfer resistance on the sulfur electrode, and C_2 interprets the double layer capacitance distributed on the surface of the pores in the sulfur electrode.

In [2], the model of a pouch Li-S cell for the calculation of the volume energy density (E_v) was elaborated (Fig. 45). In the model, the packaging materials are not considered, because the fractional volume of packaging materials is negligible in a multielectrodes pouch cell. A sandwich-structured model was used consisting of Al (7.5 μm) and Cu (4.5 μm) foils, separator (15 μm), lithium foil, and sulfur cathode. Regarding the thicknesses of the

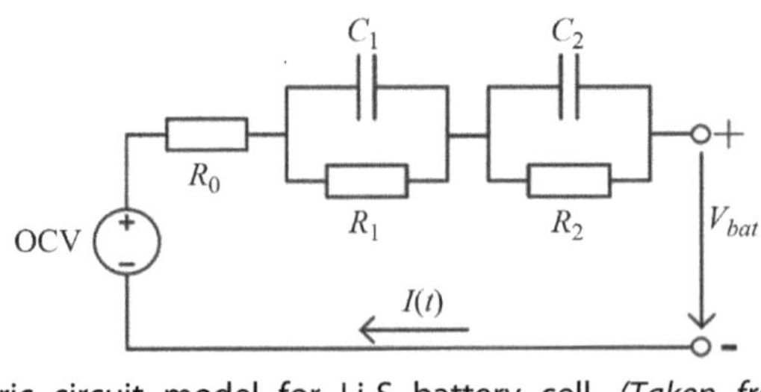

Fig. 44 An electric circuit model for Li-S battery cell. *(Taken from V. Kolosnitsyn, E. Kuzmina, et al., A study of the electrochemical processes in lithiumsulphur cells by impedance spectroscopy, J. Power Sources 196(3) (2011) 1478–1482.)*

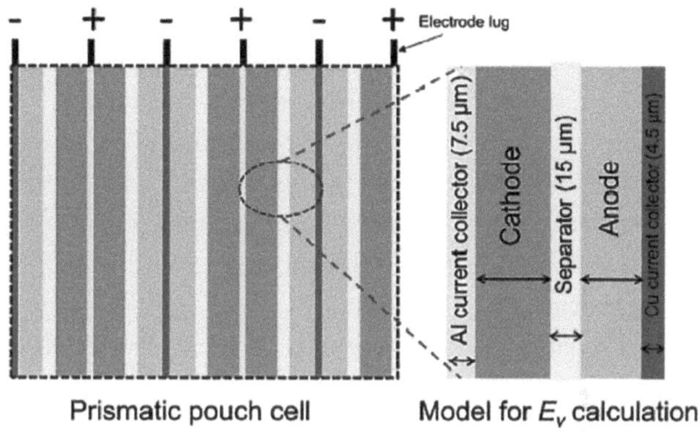

(A)

$$E_v = \frac{V \cdot m_A \cdot C_s}{(\sum m_i/\rho_i)/\varepsilon + m_l/\rho_l + \sum d_i \cdot S}$$

E_v: volumetric energy density (Wh L^{-1}); V: the average cell voltage (V);
m_A: the mass of active material (mg); C_s: the specific capacity (mAh g^{-1});
m_i: the mass of sulfur, carbon, or binder (mg); ρ_i: the density of sulfur, carbon, or binder (g cm^{-3});
ε: the porosity in the cathode; m_l: the mass of lithium (mg);
ρ_l: the density of lithium (g cm^{-3}); d_i: the thickness of Al, Cu current collectors or separator (cm)
S: unit area, 1 cm^2

(B)

Fig. 45 (A) Simplified schematic of a pouch Li-S cell and the corresponding model for the calculation of the volume energy density (E_V); (B) the equation used to calculate E_V. *(Taken from W. Xue, L. Miao, L. Qie, C. Wang, S. Li, J. Wang, J. Li, Gravimetric and volumetric energy densities of lithium-sulfur batteries, Review Article, Curr. Opin. Electrochem. 6 (2017) 92–99.)*

current collectors, half of the thicknesses of the foils was used in the model due to the widely industrially used double-side coating technology. It was assumed that the cathodes are composed of sulfur, carbon, binder (5 wt%), and interspaces between them (described as porosity (ε)). When calculating E_v, it was assumed that no excess electrolyte was used in the Li-S cells, but just that sufficient to fill the interspaces in the sulfur cathode and separator. This assumes the Li anode is fully dense and develops no porosity or thickness change of its own, which is of course overly optimistic. Based on these assumptions, E_v and E_g can be calculated using the equation shown in Fig. 1B and Supplementary Equation (1), respectively. Various parameters used for the calculation are summarized in the supplementary information.

During studies described in [14] the thermal model of a commercial 24 V, 60 Ah battery module, manufactured by Altairnano, was elaborated. The investigated module consisted of 10 LTO pouch cells connected in series. Aluminum plates were used between each two cells for the heat dissipation from the LTO pouch cell flat surfaces. All aluminum plates were welded to the module bottom, also made of aluminum. The container of the battery module was made of plastic. A phase change material (PCM) was utilized in the module to improve its thermal characteristics. The investigated battery module had three temperature sensors: on the top, in the middle, and on the bottom of the module. In addition, the voltage of each cell was measured.

Based on analysis of the heat transfer from the cells to the module bottom, top, and vertical walls, a Cauer network was used. A first-order Cauer network was used to model the heat transfer from the cells to the module bottom, because heat transfer through only one material was found. A second-order Cauer network was used to model the heat transfer from the cells to the vertical walls and to the top of the battery module. The total lumped-parameter thermal model, which was chosen to model the heat transfer and temperature in the battery module, is shown in Fig. 46.

In the mentioned lumped-parameter thermal model, the heat generated in the cells increases the average cell temperature T_{bat}. That process was modeled by an element like the current source in an electrical circuit

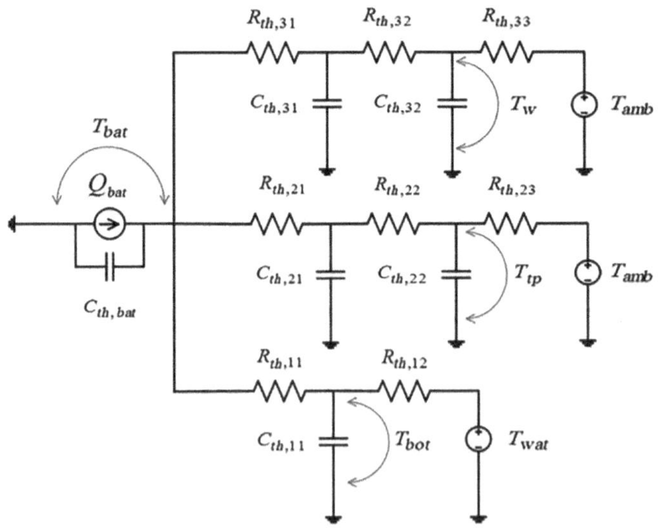

Fig. 46 Lumped-parameter thermal model of the battery module. *(Taken from K. Murashko, Thermal Modelling of Commercial Lithium-Ion Batteries, Dissertation, Lappeenranta University of Technology, Lappeenranta 2016, Acta Universitatis Lappeenrantaensis 685, ISBN 978-952-265-907-1.)*

Q_{bat} and the thermal capacitance $C_{th,\ bat}$. The heat dissipation from the cells took place in three ways. The largest amount of heat was dissipated through the aluminum bottom of the module. That process was modeled by the thermal resistance $R_{th,\ 11}$, which indicated heat transfer from the cell to the aluminum bottom. The thermal resistance $R_{th,\ 12}$ indicated heat transfer from the aluminum bottom to the water cooling system. The thermal capacitance $C_{th,\ 11}$ was used for the modeling of the module cooling system heat capacity. The element that was like the voltage source in an electrical circuit was used to model the cooling liquid temperature T_{wat}. The heat dissipation from the module top and temperature on the model top T_{tp} were modeled by the thermal resistances $R_{th,\ 21}$, $R_{th,\ 22}$ and $R_{th,\ 23}$ and the thermal capacitances $C_{th,\ 21}$ and $C_{th,\ 22}$. The heat dissipations from the vertical walls and temperature on the walls of the module T_w were modeled by the thermal resistances $R_{th,\ 31}$, $R_{th,\ 32}$ and $R_{th,\ 33}$ and the thermal capacitances $C_{th,\ 31}$ and $C_{th,\ 32}$. The ambient temperature was modeled by an element that was analogous to the voltage source in an electrical circuit, T_{amb}. The lumped-parameter thermal model parameters were estimated by approximating the measured thermal impedances of the battery module in the direction under consideration.

The detailed method for the estimation of the thermal model parameters was also presented in [14].

REFERENCES

[1] BU-301a: Types of Battery Cells, Available from:http://batteryuniversity.com/learn/article/types_of_battery_cells. Accessed 20 March 2018.

[2] W. Xue, L. Miao, L. Qie, C. Wang, S. Li, J. Wang, J. Li, Gravimetric and volumetric energy densities of lithium-sulfur batteries, Curr. Opin. Electrochem. 6 (2017) 92–99.

[3] O. Salihoglu, R. Demir-Cakan, Factors affecting the proper functioning of a 3Ah Li-S pouch cell. J. Electrochem. Soc. 164 (13) (2017) A2948–A2955, https://doi.org/10.1149/2.0271713jes.

[4] L. Qie, A. Manthiram, High-energy-density lithium–sulfur batteries based on blade-cast pure sulfur electrodes. ACS Energy Lett. 1 (1) (2016) 46–51, https://doi.org/10.1021/acsenergylett.6b00033.

[5] S. Socher, C. Jehle, U. Potthoff, Ladezustandsbestimmung von Lithium-Schwefel-Zellen mit Hilfe von in-situ Impedanzspektroskopie, Elektrik, Elektronik in Hybrid- und Elektrofahrzeugen und elektrisches Energiemanagement VII, in: C. Hoff (Ed.), On the Occasion of the 7th Conference "Electric & Electronic Systems in Hybrid and Electric Vehicles and Electrical Energy Management", Haus der Technik e.V. Essen, June 8th and 9th, 2016, in Wiesloch Renningen, 2016, pp. 213–224 ISBN: 978-3-8169-3346-5, ISBN: 3-8169-3346-7.

[6] C. Weller, S. Thieme, P. Härtel, H. Althues, S. Kaskel, Intrinsic shuttle suppression in lithium-sulfur batteries for pouch cell application, J. Electrochem. Soc. 164 (14) (2017) A3766–A3771, https://doi.org/10.1149/2.0981714jes.

[7] J. Kulisch, H. Sommer, T. Brezesinski, J. Janek, Simple cathode design for Li–S batteries: cell performance and mechanistic insights by in operando X-ray diffraction, Phys. Chem. Chem. Phys. 16 (2014) 18765.
[8] A. Fotouhi, D.J. Auger, L. O'Neil, T. Cleaver, S. Walus, Lithium-sulfur battery technology readiness and applications—a review, Energies 10 (2017) 1937, https://doi.org/10.3390/en10121937.
[9] OXIS, Energy, Available online:http://www.oxisenergy.com. Accessed 29 September 2017.
[10] S. Walus, C. Barchasz, J.-F. Colin, J.-F. Martin, E. Elkaïm, J.-C. Leprêtre, F. Alloin, New insight into working mechanism of lithium/sulfur batteries: in situ and operando X-ray diffraction characterization, Chem. Commun. 49 (2013) 7899–7901, https://doi.org/10.1039/C3CC43766C.
[11] V. Knap, D.-I. Stroe, R. Teodorescu, M. Swierczynski, T. Stanciu, Comparison of parametrization techniques for an electrical circuit model of lithium-sulfur batteries, in: 2015 IEEE 13th International Conference on Industrial Informatics (INDIN), 2015, , pp. 1278–1283.
[12] V. Kolosnitsyn, E. Kuzmina, et al., A study of the electrochemical processes in lithiumsulphur cells by impedance spectroscopy, J. Power Sources 196 (3) (2011) 1478–1482.
[13] C. Barchasz, J.-C. Leprêtre, et al., New insights into the limiting parameters of the Li/S rechargeable cell, J. Power Sources 199 (2012) 322–330.
[14] K. Murashko, Thermal Modelling of Commercial Lithium-Ion Batteries (Dissertation), Lappeenranta University of Technology, Lappeenranta, 2016. Acta Universitatis Lappeenrantaensis 685, ISBN 978-952-265-907-1.

CHAPTER ELEVEN

Thermal Batteries

As stated in Ref. [1], thermal batteries include batteries using Li- or Na-based anodes. Apart from lithium-metal-polymer batteries, thermal batteries have melted salt electrolyte and work well above ambient temperature (see Table 4).

As stated in Ref. [7], attention has been paid to the development of long-lasting batteries with molten electrolyte for high current densities and high specific powers with sufficiently high specific energies. There have been many studies on this subject. The main problem with these batteries is related to their materials, because the structural elements in these batteries must withstand aggressive factors (strong oxidants and high-temperature melting). Serious difficulties are also associated with cyclical thermal loads, i.e., changes in battery temperature from ambient temperature to operating temperature and vice versa.

A popular group of primary cells utilizes lithium as the anode and iron disulfide as the cathode. They can reach a capacity of 400 mAh g^{-1}, an

Table 4 Characteristics of Thermal Cells for Stationary and Automotive Applications [1]

System	Voltage Range (V)	Operating Temperature (°C)	Life Cycle (Cycles)	Unit Energy (Wh/kg)	Energy Density (Wh/L)	Unit Power (W/kg)
Li-Al/FeS[a]	1.7–1.0	375–500	1000	130	220	240
Li-Al/FeS_2[a]	2.0–1.5	375–450	1000	180	350	400
Na/S[b]	2.0–1.8	300–350	6000	155[c]	300[c]	250[d]
				175[d]	350[d]	
Na/$NiCl_2$[e]	2.1–1.7	250–300	2500	115	190	260
Li-metal-polymer[f]	3.0–2.0	40–60[c]	800	140[c]	174[c]	260
		60–80[d]		120[d]	160[d]	

[a]According to Refs. [2,3].
[b]According to Ref. [4].
[c]Desktop applications.
[d]Transport.
[e]According to Ref. [3].
[f]According to Refs. [5, 6].
Source: According to G. Pistoia, Nonaqueous batteries used in industrial applications, in: M. Broussely, G. Pistoia, (Eds.), Industrial Applications of Batteries: From Cars to Aerospace and Energy Storage, Elsevier, Amsterdam, 2007 (Chapter 1).

Next-Generation Batteries With Sulfur Cathodes
https://doi.org/10.1016/B978-0-12-816392-4.00011-6

energy density of 921 Wh kg^{-1}, and a voltage of 2.3 V [8]. Lithium/iron disulfide primary batteries are well known and are commercially available [9]. High-temperature lithium/iron disulfide rechargeable batteries using molten salt electrolytes have been studied [10] but have not reached production. Ambient temperature lithium/iron disulfide rechargeable batteries have also been investigated [11]. From the point of view of long cycle life and safety, a lithium-ion rechargeable battery is preferable. One difficulty in making a lithium-ion/iron disulfide battery is the need to synthesize the discharge product: lithiated iron sulfide, Li_2FeS_2. The process [12] involves heating lithium sulfide and iron sulfide at 870°C for 35 days. This is inconvenient and very wasteful of energy and is impossible for industrial production. A newer process involves heating lithium sulfide and iron sulfide for 5 h at 950°C in a vitreous carbon crucible inside an evacuated silica tube [13]. Other syntheses have been developed in which a solvent, such as a molten sulfur [14] or a molten salt [15], was used to reduce the temperature of the reaction and to speed the process. Such materials were tested in cells, like those described in Ref. [10], and in envelope (pouch) cells.

As reported in Ref. [8], lithium iron sulfide was synthesized by the reaction of lithium sulfide and iron sulfide in a solvent of molten lithium chloride. Stoichiometric amounts of reagents were mixed together and a salt, either lithium chloride, bromide, or iodide, was added. The furnace was heated to temperatures above the melting points of the added salts (lithium chloride 610°C, lithium bromide 552°C, or lithium iodide 446°C) and held at that temperature for about an hour. After cooling, the salt was removed from the lithium iron sulfide product by Soxhlet extraction by refluxing with a suitable organic solvent (pyridine for LiCl, ether for LiBr, acetonitrile for LiI) to dissolve the salt.

11.1 LI-AL/IRON SULFIDE BATTERIES

As stated in Ref. [1], these batteries contain a lithium alloy anode and a cathode made of iron sulfide, immersed in a LiCl-LiBr-KBr electrolyte. Using a dense FeS_2 cathode in this electrolyte, over 1000 cycles were obtained. The production of low-electrolyte bipolar cells or stacks is possible by using ceramic sealants from an aerobic salt. These materials create a strong bond between different metals and ceramics. Usually a separator made of pressed MgO powder is used. Electrochemical reactions for FeS and FeS_2 (only in the upper plateau) can be described as

$$2Li - Al + Fe + FeS \leftrightarrow Li_2S + 2Al \quad (17)$$

$$2Li - Al + FeS_2 \leftrightarrow Li_2FeS_2 + 2Al \tag{18}$$

Stopping discharge at the end of the upper plateau increases the reversibility of FeS_2 and extends its lifetime [16]. FeS_2 has higher sulfur activity than FeS and hence higher voltage, but it has corrosion problems. The first material is also a better electronic conductor, which makes it possible to use thicker electrodes.

The configuration of the bipolar cell, where the negative and positive electrodes are in contact with the back through the conductive plate, is more advantageous than the unipolar one. The electrodes have a large surface area and the cell structure is compact; in addition, thanks to the use of a poor electrolyte and the MgO separator, higher output power is obtained. The unit energy and power values for cells (not batteries) are shown in Table 4. As far as energy is concerned, both cells meet the United States Advanced Battery Consortium (USABC) target in the medium term (80–100 $Wh\,kg^{-1}$) and the Li/FeS_2 cell achieves a long-term target (200 $Wh\,kg^{-1}$). However, these objectives relate to fully developed batteries. The unit power of the Li/FeS_2 cell is very interesting. Indeed, the mid- and long-term values set by the United States Advanced Battery Consortium (USABC) are 150–200 and 400 $W\,kg^{-1}$, respectively. Other positive features of this system are its ability to withstand cyclic loads (over 1000 cycles) and its tolerance for overloading, excessive discharges, freezing and thawing cycles, as well as abuse resistance. The Li-Al/FeS_x system has greater safety than the sodium/sulfur; with respect to the sodium chloride/nickel system, it has a higher unit power (but less safety). As with any other thermal battery, energy is used to keep the cells warm in standby periods; in addition, the bipolar design can create production problems and high costs. This system remains a viable candidate for stationary energy storage, while tests of its traction applications have been discontinued. Finally, it should be noted that the electrolytes in the form of molten salts are displaced by polymeric electrolytes.

In Ref. [17] a rechargeable lithium aluminum/iron sulfide battery is described. Prismatic LiAl/FeS cells provided a delivered energy to 80% depth of discharge (DOD) of 217 $Wh\,L^{-1}$ and a peak power of 375 $W\,L^{-1}$. Cycle life of over 365 cycles was achieved at the C/3 rate. Pulse self-discharge and driving cycle tests were also performed for an electric vehicle application. The molten salt electrolyte required an operational temperature of about 450°C. A thermal control system composed of heating elements, thermal enclosure, and heat exchanger was engineered and tested. The components for an 8-kWh battery were developed and evaluated.

11.1.1 Anodes

As stated in Ref. [7], most of the proposed battery types used lithium anodes. In such batteries, the electrolyte must contain a lithium salt, and electrode processes on the lithium anode consist of a simple transfer of lithium ions from the crystal lattice of the metal to the molten material and vice versa. At the working temperature of the battery (400–600°C), pure lithium is a liquid. There were two main types of lithium electrode designs: liquid lithium in a porous matrix and solid lithium alloyed with a different metal.

The lithium-containing electrode could be made of porous stainless-steel plates, felt-like stainless steel, nickel, or similar material, having a high porosity (up to 90%), with a small pore diameter and the required elasticity.

The polarity of the lithium electrode is negligible during both loading and unloading; for lithium electrodes, record current densities have been obtained, up to $40\,A/cm^2$.

One of the main disadvantages of liquid lithium is its noticeable solubility in molten salt, and its ability to expel potassium from the alloy:

$$Li + KCI \rightarrow K + LiCl \tag{19}$$

The potassium vapor pressure is much higher than the lithium vapor and, moreover, the potassium can form an undesired gas phase. The lithium dissolved in the electrolyte travels towards the cathode when it is consumed in a nonproductive chemical reaction. For an interelectrode distance of 1 mm, the associated self-discharge is equivalent to a leakage current density of $1\text{–}10\,mA\,cm^{-2}$. In addition, dissolved lithium causes the breakdown of ceramic separators. The solubility of lithium increases significantly with increasing temperature. Lithium alloys can replace lithium to reduce activity and solubility in the electrolyte. Lithium alloy electrodes that are liquid at operating temperature (zinc, tin, etc. alloys) have the same construction as pure lithium electrodes. Solid lithium alloys (with silicon or aluminum) are used in the form of porous pressed panels on electrically conductive nets. Substitution of pure lithium by the alloys results in a reduction of the opening voltage (OCV) and operating voltage by 0.2–0.3 V, but this is justified by a significant reduction in self-discharge and a longer period of use.

As reported in Ref. [1], Li-Al/iron sulfide batteries contain a lithium alloy anode that is multiphase $(\alpha+\beta)$ Li-Al and $Li_5Al_5Fe_2$.

11.1.2 Cathodes

As stated in Ref. [7], the first prototypes of molten electrolyte batteries used cathodes with chlorine introduced by gas diffusion, like the electrodes of fuel cells. Even the early prototypes of batteries had discharge current densities of up to $4\,A\,cm^{-2}$ without significant polarization. The chlorine electrode is made of porous graphite, boron, and silicon carbides or similar materials. The difficulty associated with the development of these electrodes consisted in the selection of a molecular chlorine storage technique. Originally, the use of activated carbon for chlorine adsorption has been suggested. Coal, with a more developed real surface, could provide chlorine adsorption capacity up to $0.3–0.5\,Ah\,cm^{-3}$. Another suggestion was to store chlorine in vessels under elevated pressure. However, none of these proposals has been implemented.

Later, there appeared a proposal to use sulfur, and then aerobes, mainly sulfides, as active materials for the cathode. Sulfur is liquid at the temperature of the battery. The sulfur cathode was produced from a mixture of sulfur and carbon dioxide, or in the form of a niobium box with a niobium filler packed with sulfur. The high volatility of sulfur (at 507°C the sulfur vapor pressure is 2 atm), and its solubility in the molten electrolyte led to self-discharge. The greatest success was obtained for iron sulfides, FeS and FeS_2. Both sulfides have a high Ah capacity and are cheap and nontoxic. The electrode production process is simple [18].

The discharge and charging processes in sulfide electrodes can be described as follows:

$$FeS + 2e^- \underset{ch}{\overset{disch}{\rightleftarrows}} Fe + S^{2-} \tag{20}$$

and

$$FeS_2 + 4e^- \underset{ch}{\overset{disch}{\rightleftarrows}} Fe + 2S^{2-} \tag{21}$$

The reaction with FeS_2 involves the formation of an intermediate FeS; therefore the charge-discharge curve of the battery with the FeS_2 electrodes consists of two gently sloping parts with approximately the equivalent voltage length of 2.05 and 1.65 V. Theoretical unit consumption of FeS_2 is slightly smaller than FeS (11.2 and 1.64 g/Ah) but FeS_2 has higher corrosion activity, which may lead to a shorter service life. The sulfide electrodes are made of a mixture of FeS or FeS_2 with some additives (lithium, copper, and cobalt sulfides) placed in a porous molybdenum cage, tungsten, graphite, and so on, and the electrolyte is added to the cathode. For example, a cathode may have the following composition: 60% FeS + 2.2% Li_2S + 29.3% LiCl + KCI eutectic + 7% carbon + 1.5% iron powder (in percent by mass). Iron is added to prevent the formation of elemental sulfur in the event of overload: it reacts with Li_2S to form FeS_2.

There were relatively high current densities (up to $0.4\,A\,cm^{-2}$) using sulfide cathodes.

As reported in Ref. [1], Li-Al/iron sulfide batteries contain a cathode of iron sulfide. Both FeS and FeS_2 can be used as the material for this cathode.

According to Ref. [16], in Li-Al type/iron sulfide accumulators, cathodes are produced by compressing a mixture of FeS_x and electrolyte on a current collector or packing the material in a honeycomb matrix. Graphite may be included, and CoS_2 and NiS_2 are also sometimes added.

11.1.3 Electrolyte

As stated in Ref. [7], the maximum specific capacities of lithium-chlorine and lithium-sulfide batteries can be obtained for electrolytes containing pure LiCl or Li_2S, which are products of a current-producing reaction. However, their melting points are too high (613°C and 950°C, respectively), therefore mixtures of LiCl + KC and LiCl + KCI + Li_2S with lower melting temperatures are usually used as electrolytes to obtain a working temperature no higher than 400°C (the melting point of the LiCl + KCI eutectic is 352°C). The presence of inert additives in the electrolyte results in a certain decrease in the specific capacity of the battery. In addition, the electrolyte composition changes during operation, because the content of LiCl or Li_2S increases with discharge and decreases with charging, and at the same time the melting temperature of the electrolyte changes.

Most types of batteries use immobilized or matrix electrolytes. The fine boron nitride powder, lithium aluminate, etc., are used as immobilizing agents, and the matrices are manufactured from ceramic materials such as boron nitride, stabilized zirconium oxide, and so on.

As reported in Ref. [1], Li-Al/iron sulfide batteries contain electrolyte; this is a LiCl-LiBr-KBr (high in LiCl) low-impedance mixture.

11.1.4 Prototypes

As reported in Ref. [7], large-scale research and development projects on lithium-FeS and Li-FeS_2 batteries are underway at the Argonne National

Laboratory and several other US organizations. Two types of batteries have been developed, one for electric vehicles and military applications, and the other for load balancing in power plants. In the first type of battery, FeS_2 cathodes were used, ensuring higher efficiency (nominal specific energy was 150 Wh kg^{-1}), but shorter service life and planned life (planned life was 1000 cycles, and lifetime 3 years) and the need to use more expensive materials. The second type of battery used FeS cathodes. The specific energy was half that for FeS_2 batteries, but their cost was lower by almost 50%, and the planned life and service life were up to 3000 cycles and 10 years, respectively. The second type of battery was designed for lower charging and discharging currents and had higher efficiency. When the batteries did not work they could be cooled to room temperature—that is, the electrolyte could be frozen. The batteries could then be heated to operating temperature without losing capacity or lowering the parameters. This significantly simplified the long-term storage of batteries. One of the battery prototypes for electric vehicles had a volume of 320 L and a mass of 820 kg. The cathode was made of FeS with the addition of CoS_2. Several layers of active material alternating with the graphitized material were placed in a basket of molybdenum mesh welded to the central molybdenum current collector. The cathode was wrapped by a two-layer separator. The inner layer consisted of a ZrO_2 material and the outer layer was made of a boron nitride (BN) fabric. The anode consisted of a lithium alloy with silicon in a porous nickel matrix. The container and cover were made of stainless steel, electrically connected to the anode. The prototype was loaded with current up to 50 A, and the specific power was up to 53 W kg^{-1} [19]. A larger battery, intended for a submarine, had a closed container with six cathodes and six anodes. The anodes were made of a lithium-aluminum alloy. Separators made of BN fabric were placed between the electrodes. The battery had the following dimensions: diameter 30.5 cm; height 21.1 cm; weight 43 kg; rated discharging voltage 1.45 V; and the correct capacity is about 150 Wh kg^{-1}. The battery was designed for a normalized current consumption of $j_d = 0.08$.

11.2 LI-CoS_2 BATTERIES

According to Ref. [20], the primary electrochemical system used in current thermal batteries is the Li(Si)/FeS_2 couple with a nominal output voltage of 1.94 V at 500°C. However, for special higher-power applications, the FeS_2 is replaced by CoS_2, which is made synthetically and is more expensive.

According to Ref. [20] this cell with a CoS_2 cathode, LiBr-KBr-LiF electrolyte, and Li(Al) anode combination can obtain high specific energy and extended periods of open circuit without loss of capacity.

During comparative studies described in Ref. [21], the Li(Si)/FeS_2 and Li(Si)/CoS_2 couples were evaluated with a low-melting LiBr-KBr-LiF eutectic and all-Li LiCl-LiBr-LiF electrolyte for a battery requiring both high energy and high power for short duration. Studies were carried out with 1.25-in. diameter triple cells and with 10-cell batteries. The Li(Si)/LiCl-LiBr-LiF/CoS_2 couple performed the best under the power load and the Li(Si)/LiCl-LiBr-LiF/FeS_2 was better under the energy load. The former system was the best overall performer for the wide range of temperatures for both loads, because of the higher thermal stability of CoS_2.

Also, in Ref. [22] an evaluation of Li(Si)/LiCl-LiBr-LiF/CoS_2 and Li (Si)/LiCl-LiBr-LiF/FeS_2 electrochemical systems was carried out. The internal resistance of batteries based on a cobalt disulfide (CoS_2) cathode was 24% less than that based on an iron disulfide (FeS_2) cathode; the peak voltage in parallel of 17 cells based on the CoS_2 cathode was 32.3 V, but that based on the CoS_2 cathode was 33.5 V. The performance comparison of thermal batteries based on the CoS_2 cathode was evaluated at 25°C, −47°C, and +62°C, respectively. The discharge curves at different temperatures almost overlapped.

According to Ref. [23], CoS_2 with a high decomposition temperature of 620°C has low open-circuit voltage and high cost.

According to Ref. [24], a hydrothermal method for the preparation of CoS_2 cathodes with pyrolytic carbon was developed. The CoS_2/C cathodes had excellent discharge performances in thermal batteries because carbon coating improved the decomposition temperature of nano-CoS_2 (410°C rises to 610°C). The hierarchical carbon modification was designed to improve the thermal stability and to ensure high specific capacity and long life of thermal batteries. However, carbon effects have not been clearly studied in thermal batteries.

One modification of the cathode made of CoS_2 is the B/CoS_2 cathode. According to Ref. [25], the Li-B/CoS_2 couple was designed. Its capacity and safe properties depended on the acute self-discharge that resulted from the dissolved lithium anode in molten salt electrolyte. To solve those problems, carbon-coated CoS_2 was prepared by a pyrolysis reaction of sucrose at 400°C. The carbon coating as a physical barrier protected CoS_2 particles from damage by dissolved lithium and reduced the self-discharge reaction. Therefore, both the discharge efficiency and safety of Li-B/CoS_2 thermal batteries increased. Discharge results showed that the specific capacity of the first discharge plateau of carbon-coated CoS_2 was 243 mAh g^{-1}, which was 50 mAh g^{-1} higher than that of pristine CoS_2 at a current density of 100 mA cm^{-2}. The specific capacity of the first discharge plateau at 500 mA cm^{-2} for carbon-coated CoS_2 and pristine CoS_2 were 283 and 258 mAh g^{-1}, respectively. The carbonization process did not influence the intrinsic crystal structure and thermal stability of pristine CoS_2.

11.3 LI-NiS_2 BATTERIES

NiS_2 is a cathode material found in primary batteries that operate at high temperature. In Ref. [26] the in situ battery discharge study of a thermal battery cell is presented. To obtain this cell, $Li_{13}Si_4$ (anode, 0.22 g, 1.1 mmol), a mixture of LiCl-KCl eutectic (electrolyte) and MgO (separator) and NiS_2 (cathode, 0.44 g, 3.6 mmol) were pressed into pellets of diameter 23.6 mm. These pellets were kept under argon before loading into a Swagelok type assembly in an argon-filled glovebox. This enabled electrochemical testing to be carried out ex situ at 520°C in a muffle furnace, using a Maccor battery tester (model 5300). The battery was tested galvanostatically, applying a current density of 22.9 mA cm^{-2}. Five different regions were observed upon battery discharge and four different nickel-containing phases occurred (NiS_2, NiS, Ni_3S_2, and Ni). A new discharge mechanism was proposed that did not include Ni_7S_6.

NiS_2 with its low cost has been widely considered as a cathode material because its thermal stability and discharge performances are intermediate between those of FeS_2 and CoS_2 [26, 27]. However, the high specific capacity is a significant problem for NiS_2 cathodes in thermal batteries, because the NiS_2 cathode cannot completely react at high current density in application. Nanocrystallization is an effective method to improve the specific capacity of the cathode by increasing the specific surface area, shortening the transmission distance between Li^+ and electrons [28–34].

Nanocrystallization indeed accelerates intermediate phase evolution and obtains a high specific capacity of the NiS_2 cathode in a thermal battery [35]. However, due to the high defect density and grain boundary volume of nanocrystallization, the decomposition temperature of transition metal sulfides decreases [36–38]. This characteristic makes it only apply to short-life thermal batteries (discharge temperature $\leq$500°C).

According to Ref. [39], nano-NiS_2 powders were successfully synthesized by a simple low-temperature, solid-state method ($Ni + S \rightarrow Ni_3S_2 + S \rightarrow NiS_2$) [35]. Nanocrystallization is widely used to improve the discharge performances of LIBs. While its lower thermal stability limits the operating time, especially for the thermal battery system, which discharges at high temperatures (usually 500–550°C). As described in Ref. [39], NiS_2 particles were coated with amorphous carbon. Then they accumulated into submicron particles and were connected/fixed by a carbon network. The initial decomposition temperature of nano-NiS_2 increased from 400°C to 590°C after this hierarchical carbon modification. The hierarchical carbon modified nano-NiS_2 cathode revealed high discharge performances in thermal batteries at high temperatures. Specifically, with $0.1\,A\,cm^{-2}$ at 500°C, the specific capacity and energy reached $610\,mAh\,g^{-1}$ and $1082\,Wh\,kg^{-1}$, respectively, at a cut-off voltage of 1.4 V (the minimum operating voltage = the maximum voltage × 0.7), which was higher than those of similar state-of-the-art sulfide-cathode materials for thermal batteries in the literature. The specific energy retained about $503\,Wh\,kg^{-1}$ even at 700°C. The multiple protective effects of hierarchical carbon modification improved the conductivity and thermal stability and inhibited the dissolution and shuttling of products. So, hierarchical carbon modification made nano-NiS_2 suitable for high specific energy and long operating life thermal batteries.

11.4 SULFUR-SODIUM BATTERIES

As stated in Ref. [40], the exemplary Na-S battery was developed by Tokyo Electric Power Company (TEPCO) and NGK Insulators Co., Ltd. since 1983 [41]. Studies of various prototypes showed that the Na-S cell technology is attractive for use in relatively large energy storage systems due to its outstanding energy density, efficiency, no maintenance, and long life of up to 15 years [41–44]. Currently, the development of Na-S cells is focused on multifunctional energy storage systems, performing several energy management functions, and nonstandard power devices that improve the quality of energy supplied. The efficiency of the Na-S cell is the main factor in determining its suitability for various applications. The features and advantages of Na-S cells are discussed in Refs. [41, 42, 45, 46].

According to Ref. [47], the concepts used for Li-S batteries can be transferred to sodium coupled with an elemental sulfur cathode. The main advantage of developing a room-temperature Na-S battery is to further reduce the battery cost by using the abundant elemental sodium available in nature and avoiding the possible safety hazard of high-temperature Na-S batteries. However, this battery system faces various challenges, many of which are like those of its Li-S counterparts (e.g., an even larger cathode charge-discharge volume change and similar shuttling of Na polysulfide intermediates). This makes the design principles of S-cathodes in a Li-S battery partially applicable to a room temperature Na-S battery.

As reported in Ref. [7], the development of high-temperature sodium sulfide batteries using solid sodium polyaluminate as an electrolyte was first described in 1966 by the Ford Company in the United States. At the time, such batteries were essential for electric vehicles. Currently, they are only predicted for unknown yet possible future applications.

The main advantages of these batteries are high specific power and energy, good reversibility, no side reactions, hermetic sealing, and cheapness and free availability of the main reactants, sulfur and metallic sodium. The disadvantage of the batteries is the high working temperature, in the range of 300–350°C. Unlike other types of batteries, these involve solid electrolyte and liquid reagents not only through mechanical stress, but also heterogeneity of the electrolyte structure and the presence of defects. Electrolyte durability is significantly improved thanks to careful regulation of the electrolyte microstructure production and control. An important factor in this respect is the homogeneity of the particle size of the original electrolyte; the optimal size is 2–4 μm.

As stated in Ref. [1], high-temperature batteries using sodium as a negative electrode, sulfur as positive, and β''-Al_2O_3 as a solid Na^+ electrolyte are currently used for energy storage applications. The energy density of the Na/S system is very high (see Table 4) and significantly exceeds that of the water (solution) (lead-acid and nickel-cadmium) systems previously used for energy storage. Other distinct advantages of Na/S batteries are: good power density, long life cycle, regardless of the outside temperature, and the operating temperature of the system is between 300 and 350°C. In this temperature range, both Na and S are liquid, and the solid electrolyte has a high Na^+ ion conductivity, thus providing good kinetics. During discharge, Na^+ ions migrate from Na to S and form polysulfides, the formation of Na_2S_3 at 1.78 V is regarded as a limitation at discharging. At C/3, the average voltage is 1.9 V.

During charging, reactions leading to Na_2S_3 are reversed, and in the final stage there is a significant increase in the resistance due to the nature of the insulating sulfur. Thus, the charging must be stopped before the total recovery of Na, and subsequent discharges provide 85%–90% of the theoretical capacity [4]. This system has a high capacity for cyclic loads (up to 5000–6000 cycles). This is mainly due to the liquid state of reagents and products, as the aging mechanism based on morphological changes in electrodes does not work here.

Interesting information was given in Ref. [48], that fine cathode materials Ni_3S_2 and NiS_2 were synthesized using the simple, convenient process of mechanical alloying (MA). To improve the cell properties, wet milling processes were conducted using low-energy ball milling to decrease the mean particle size of both materials. The cells of Na/Ni_3S_2 and Na/NiS_2 showed a high initial discharge capacity of 425 and 577 $mAh\,g^{-1}$, respectively using wet milled powder particles, which is much larger than commercial ones, providing some potential as new cathode materials for rechargeable sodium-ion batteries.

11.4.1 The Design of the Na-S Battery and Its Cathode

As reported in Ref. [7], sulfur-sodium batteries have a mainly tubular structure with a tube-shaped electrolyte 20–50 cm long, 1.5–3.5 cm in diameter, and a wall thickness of about 1 mm. One reagent is inside the tube and the other outside. Molten sulfur and polysulfides are usually found inside the tube. The molten sodium is in the gap between the electrolyte and the wall of the battery tank. The sodium resource is stored in a container at the top and bottom of the battery.

Because both sulfur and polysulfides lose electron conductivity, very loose felt graphite material is used as a current cathode collector: its mass is only 3%–10% of the mass of reagents. Felt graphite also has a different function; during charging (at 14.2 V) sulfur is formed on the electrolyte surface. Because sulfur loses ionic conductivity (as opposed to polysulfides), it can block polysulfide melting, which makes it difficult to continue charging. Fumed graphite has good wettability through molten sulfur, but not through a molten polysulfide. When the high pore felt layer is in the vicinity of the electrolyte and the fine pore layer is further away from it, the capillary forces tend to draw the molten sulfur away from the electrolyte surface, thereby reducing the blocking effect of sulfur. This system contributes to a significant increase in the charging capacity of the sulfur electrode (at 14.2 V) in such a way that the sulfur utilization rate increases up to 80%–90%.

As stated in Ref. [1], the sulfur electrode has a surface impregnated with a layer of carbon or felt graphite. Carbon fibers provide good electronic conductivity because sulfur is an insulator for both electrons and ions. Fortunately, Na polysulfides are good ion conductors.

In Ref. [49] a flexible carbon-cloth cathode material for Na-S batteries, obtained by simply carbonizing cotton cloth in inert atmospheres, was reported. Such freestanding and flexible carbon cloth acted as a conductive host for sulfur and resulted in an initial discharge capacity of 390 mAh g^{-1} at 0.1 *C* (1 *C*=1675 mA g^{-1}), which remained at 120 mAh g^{-1} after 300 cycles. A flexible prototype room-temperature Na-S pouch cell was then assembled using this carbon cloth as a proof of concept, and this retained a stable capacity on bending.

In Ref. [50] a microporous carbon cloth fabricated using electrospinning technology was reported for a Li-S or Na-S battery. The sulfur-selenium (S/Se) mixture, rather than pure sulfur, was used as the cathode active material and impregnated into the pores of the carbon cloth. By doing so, a high reversible capacity of 840 mAh g^{-1} after 100 cycles at 0.1 A g^{-1} for Li-S batteries, was achieved. Even at extremely high rates up to 10 and 20 A g^{-1}, high reversible capacities of 350 and 181 mAh g^{-1} were still obtained. When used as the cathode material for room-temperature Na-S batteries, it again exhibited very good cycle stability (762 mAh g^{-1} after 100 cycles at 0.1 A g^{-1}) and excellent rate capability (190 mAh g^{-1} at 2 A g^{-1}).

The Na-S battery presented in Ref. [51] utilized a Na metal anode, a metal-organic framework (MOF)-derived microporous carbon polyhedron-sulfur composite (MCPS) cathode, and a liquid electrolyte composed of a 1:1 mixture of ethylene carbonate (EC) and propylene carbonate (PC) containing 1 M $NaClO_4$ salt and 1-methyl-3-propylimidazolium-chlorate ionic liquid tethered silica nanoparticle (SiO_2-IL-ClO_4) additives as an agent for stabilizing electrodeposition. Such Na-S cells achieved cycling performance with nearly 100% coulombic efficiency at higher current density and with relatively high sulfur loadings in the cathode. Reversible storage capacities of over 860 mAh g^{-1} at 0.1 *C* (1 *C*=1675 mA g^{-1}) and 600 mAh g^{-1} at 0.5 *C* based on active sulfur mass were obtained. Even at the higher current density (0.5 *C*) the batteries possessed cycle stability for over 100 cycles with 0.31% capacity decay per cycle. No soluble NaPS species were formed and the diffusivity of Na^+ into the composite cathode was consistent with expectations for solid-state transport. The Na-S cells followed a different electrochemical reaction mechanism compared to traditional metal-sulfur batteries, which contributes to the stability and high capacity retention on cycling.

It was found that the particles formed a sodium-ion conductive film on the anode, which stabilized deposition of sodium. Sulfur remains interred in the carbon pores and undergoes solid-state electrochemical reactions with sodium ions.

11.4.2 Na-S Battery Anode

As stated in Ref. [1], an important initial requirement for the Na electrode is high purity, since other metals and Na compounds are not accepted. Pollutants concentrate on contact with the electrolyte, which reduces the active surface of the electrode or even causes damage [4].

The Na-S battery described in Ref. [51] utilized a Na metal anode.

11.4.3 Na-S Battery Electrolyte

In the sulfur-sodium batteries discussed in Ref. [7], the electrolyte is a non-barrier ceramic material shaped as a thin disk (for flat batteries) or a tube with one closed end (like a test tube) used in cylindrical batteries. The electrolyte production process consists of the following stages:

- **(i)** initial calculations and grinding of starting materials (aluminum oxide, α-Al_2O_3 and sodium carbonate);
- **(ii)** thorough mixing of the components with the binder;
- **(iii)** isostatic compaction of the powder at a pressure of about 400 MPa;
- **(iv)** sintering at a temperature of about 1600°C. Sintering is the most important step, since Na_2O sodium is the volatile substance of the sintered electrolyte in the dispensed platinum crucibles.

In general, the technological process is rather complicated, and its effectiveness, so far, is not higher than 50%–60%.

As reported in Ref. [1], the electrolyte β''-Al_2O_3 has a slight electronic conductivity and is impermeable to molten sodium and S. The idealized composition for β''-Al_2O_3 is Na_2O*5.33Al_2O_3. Pure β''-Al_2O_3 is not easy to produce, and therefore must be stabilized with Mg or Li ions, which replace Al ions. The ionic conductivity of this electrolyte is $\sim 0.5\,\Omega^{-1}\,cm^{-1}$ at 350°C for the polycrystalline form. β''-Al_2O_3 is rather sensitive to moisture, which contributes to the deterioration of its mechanical properties. Therefore, some β-Al_2O_3 (idealized formula, Na_2O*11Al_2O_3) is contained in the mixture, despite its lower conductivity, as it is less hygroscopic. Conductivity $\sim 0.2\,\Omega^{-1}\,cm^{-1}$ is considered acceptable for practical electrolyte applications [16]. The production of Na/S batteries for stationary applications is particularly active in Japan. Very large energy storage systems can

Table 5 Technical Data for the Na-S Modular Battery [44]

Output power	52.1 kW
Voltage	58 V/116 V
Current	726 A/363 A
Capacity	375 kWh
Efficiency	>83%
Dimensions	2.17 m × 1.69 m × 0.64 m
Weight	3500 kg
Energy density	160 kWh/ms
Number of cells	320

be built from Na/S modules: their capacity can be up to 57 MWh. They can live in the present application for up to 15 years and withstand thousands of cycles.

The Na-S battery presented in Ref. [51] utilized a liquid electrolyte composed of a 1:1 mixture of ethylene carbonate (EC) and propylene carbonate (PC) containing 1 M $NaClO_4$ salt and 1-methyl-3-propylimidazolium-chlorate ionic liquid tethered silica nanoparticle (SiO_2-IL-ClO_4) additives as an agent for stabilizing electrodeposition.

11.4.4 Na-S Battery Configuration

As stated in Ref. [40], Na-S batteries are usually used in a modular form, in which the cells are collected and placed in a thermal casing. The experimental module with cells was equipped with an electric heater, which had the task of raising and maintaining the temperature of the cells to the value corresponding to their optimal work, i.e., almost 300°C. Cells inside the experimental module were positioned by filling and solidifying with dry sand to prevent possible fires. Table 5 presents data for the Na-S prototype modular battery [40].

11.4.5 Factors Affecting the Operation of the Na-S Battery

Na-S battery operation is affected by several factors, in particular: internal electrical resistance, operation, temperature, electromotive force, and discharge depth [40].

The internal resistance of the Na-S battery consists of the electrical resistance associated with the electrolyte, plate, and fluid resistances, as well as the resistance related to the polarization effect. The internal resistance changes during the loading and unloading operation depending on the discharge depth and temperature, as shown in Fig. 47. One group of curves

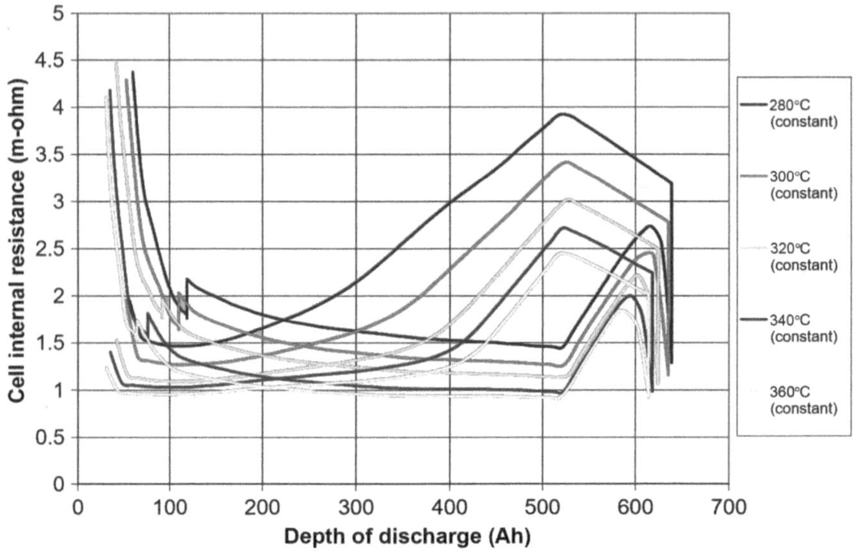

Fig. 47 Depth of discharge as a function of cell internal resistance, for different temperature values [40].

corresponds to the charging process and the other group to the discharge process for five different cell temperature values. The mentioned curves illustrate how the internal resistance depends on the depth of discharge and the various temperature values of the cells. The internal resistance decreases when the temperature rises from 280°C to 360°C. At the end of the charging and discharging operation, a polarization effect increases the internal resistance. These discharge depth ranges must be avoided when the Na-S battery is running, due to an excessive increase in its internal resistance.

Internal resistance also changes during long-term cyclic operation and depends on the number of cycles of battery charging/discharging. As shown in Fig. 48, the internal resistance increases as the number of charging and discharging cycles increases. This factor is important because it determines the remaining available peak power and output voltage of the Na-S battery.

The next factor is the working temperature of the Na-S battery, which is usually from about 300°C to 360°C. This temperature varies in different ways, between the states of charging, waiting, and discharging. During the unloading state, the Na-S accumulator generates heat caused by electrical resistance and the heat of entropy, which leads to accumulating heat in the accumulator and raising its temperature. During the charging state, the amount of thermal energy induced by electrical resistance is almost equal to

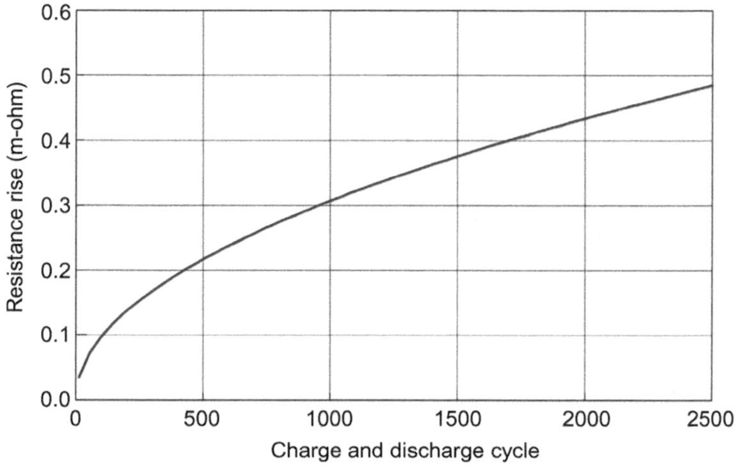

Fig. 48 Increase of cell resistance from the function of number of cycles loading/unloading [40].

the absorption of entropy heat. Thus, the temperature of the battery is gradually lowered. The heat stored in the battery is dissipated only during standby. As a result, the temperature of the battery gradually decreases. When the temperature exceeds the lower limit (300°C), the heater installed inside the modular battery is turned on to raise the temperature and keep it within the nominal range. The resistance of the Na-S battery depends on the temperature, and the higher the temperature of the module, the lower its internal resistance.

The influence of temperature on the internal resistance is very important, because it determines the limit of peak battery power. In some applications, the Na-S batteries are subjected to peak power equal to four to five times the rated power [41, 42, 45]. Peak output power at higher current generates more joule heat through internal resistance.

For example, a 50 kW module battery load lasting 30 s, corresponding to its five times rated power, causes a temperature increase of approximately 3°C [45]. During such loads with peak power, the temperature should be kept within the range corresponding to the normal working condition, to avoid reaching an unacceptably high temperature and creating undesirable temperature differences in the battery.

The next factor is the electromotive force (EMF) of the Na-S battery, which depends mainly on the depth of discharge. Because of the electrochemical reaction of the components, the electromotive force EMF of the Na-S battery is relatively constant but decreases linearly when the battery reaches 60%–75% of the discharge depth, as shown in Fig. 49.

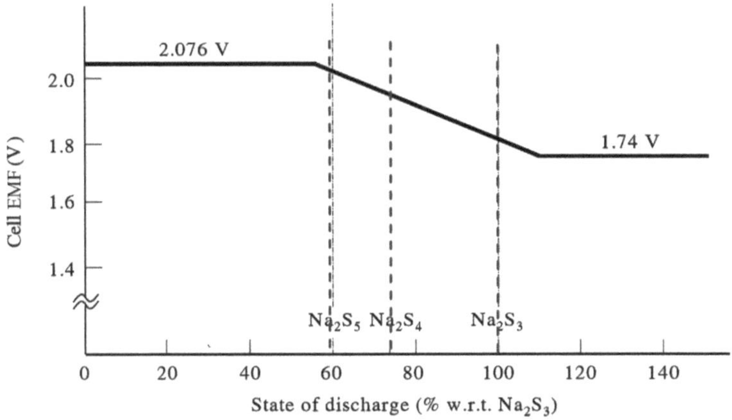

Fig. 49 Electromotive force of the Na-S cell as a function of discharge depth [40].

In practice, discharging the Na-S battery is limited to less than 100% of its theoretical capacity, due to the more corrosive properties of Na_2S_3. Before all materials are exchanged in Na_2S_3, the sodium in the cell will move to the active electrode and the sodium area will become empty. In this case, there is no path for the electron on the negative electrode, which results in low efficiency during discharge.

Hence, the battery is designed to stop discharging before all sodium passes to the active electrode. To ensure an adequate safety margin, the amount of sodium remaining in each storage cell must be considered. For example, for a total sodium volume of 780 g, the usable volume of sodium is 675 g, which makes the remaining sodium volume 13.5%.

Therefore the Na-S cells usually provide 85%–90% of their theoretical capacity, which means that the approximate composition of the sodium polysulfide accompanying the 1.82-V voltage on the cell corresponds to the mixture of Na_2S_4 and Na_2S_2 at the discharge end.

This factor is important because it allows the voltage level to be evaluated at the end of discharge and to predict the possible maximum Na-S battery power for each discharge depth.

Another factor is the discharge depth (DOD), representing the remaining capacity of the battery. It is important because it is related to changes in internal resistance, temperature, and EMF battery electromotive force.

According to Ref. [52], for the Na-S battery to work with the internal combustion engine, it must be possible to heat it up to 300°C. From the point of view of efficiency of the car, it should not require additional heating elements to obtain this temperature. The only potential source of this temperature can be hot exhaust from the exhaust system of the internal combustion engine. A thermal chamber containing such a heated battery would have to be placed as close as possible to the engine to make the best use of the high temperature of the exhaust gas. In order not to cause unnecessary increase of the exhaust flow resistance, the design of the Na-S battery would thus have to be adjusted to the shape of the engine exhaust system. The battery cell components should be placed as close as possible and along the exhaust manifold channels. Since such a battery can carry relatively large charging currents, it could thus be used as a device for accumulating recovered energy, for example, in the braking process. The use of energy accumulated in the battery would have to be accomplished through a voltage divider, limiting its value to a safe level for the vehicle's electrical circuits. The large dimensions of existing prototypes indicate such batteries must be used in larger vehicles, for example in buses.

11.5 AL-S BATTERIES

In Ref. [53] an Al-S battery based on concentrated polysulfide catholytes and an alkaline aluminum anode was presented. This battery was expressed by aluminum oxidation and aqueous sulfur reduction for an overall battery discharge consisting of

$$2Al + S_4^{2-} + 2OH^- + 4H_2O \rightarrow 2Al(OH)_3 + 4HS^- \quad E_{cell} = 1.79\ V \qquad (22)$$

The theoretical energy density of the Al/S battery (based on potassium salts) was 647 Wh kg^{-1}. An Al-S battery was demonstrated with an open-circuit voltage of 1.3 V, and energy densities based, respectively, on dry and total battery materials were 170 and 110 Wh g^{-1}.

Also, in Ref. [54] a similar battery was presented containing an elemental sulfur cathode with a measured capacity of over 900 Ah kg^{-1}, more than 90% of the theoretical storage capacity of solid sulfur at room temperature, accessed by means of a lightweight, highly conductive, aqueous polysulfide interface through the electrocatalyzed reaction $S + H_2O + 2e^- \rightarrow HS^- + OH^-$. The mentioned solid sulfur cathode mated with an aluminum anode for an overall discharge reaction $2Al + 3S + 3OH^- + 3H_2O \rightarrow 2Al(OH)_3 + 3HS^-$, giving a cell potential of 1.3 V. The theoretical specific energy of the Al-S battery (based on potassium salts) was 910 Wh kg^{-1} with an experimental specific energy of up to 220 Wh kg^{-1}.

It was explained in Ref. [55] that conventional cathodes utilized in nonaqueous lithium anode batteries, including molybdenum or titanium sulfides, are incompatible with nonaqueous aluminum anode electrochemical storage processes. Alternative cathode systems were explored to develop high charge capacity nonaqueous aluminum cells. A series of high storage capacity Al cells, utilizing cathodes composed of CF_x fluorinated polymer graphite compounds were tested. High CF_x cathodic capacities were obtained in aluminum anode cells utilizing a 0.3 M tetraethylammonium chloride, 10 mM Hg $(CH_3COO)_2$ acetonitrile electrolyte. The addition of hydroxide to the CF_x cathode mixture increased discharge potential, while the addition of nonfluorinated 1 μm graphite increased measured cathodic capacity. The CF_x cathode capacity increased with the degree of fluorination from 27% to 35% fluorine, was almost constant from 35% to 58% fluorine, increased in capacity for 61% fluorine, and increased again with 63% fluorine (the highest level of 63% fluorine available in the CF_x). Measured cathode capacities exceeded 250 $mAh\,g^{-1}$. In excess of 300 $mAh\,g^{-1}$ total cathode mass was measured for a 50:50 (wt%) cathode composite containing 55% fluorinated CF_x and the rest in the form of 1 μm nonfluorinated graphite.

In Ref. [56] an Al-S type battery consisting of a composite sulfur cathode, aluminum anode, and electrolyte in the form of ionic liquid $AlCl_3$/1-ethyl-3-methylimidazolium chloride is described. The electrochemical reduction of elemental sulfur in various molar ratios of electrolyte was investigated, and tetrachloride ions were identified in electroactive ionic substances. The Al-S battery showed a plateau of discharging voltage of 1.1–1.2 V, with a very high charging capacity of more than 1500 $mAh\,g^{-1}$, in relation to the mass of sulfur in the cathode. The energy density of the Al-S cell is estimated at 1700 $Wh\,kg^{-1}$ of sulfur, which is competitive in relation to the most attractive chemical batteries designed to store high electrochemical energy. The complete dissolution of sulfur-based discharge products in the electrolyte was noted.

In Ref. [57] it was found that metallic aluminum is a promising material for anodes in the new generation of batteries due to the prevalence of occurrence, potentially nondendritic deposition, and high capacity. Al-S batteries are interesting due to high energy density (1340 $Wh\,kg^{-1}$) and low cost. However, the drawback of Al/S chemistry is poor reversibility due to difficulties in AlS_x oxidation. During the tests, a reversible Al-S battery was used with an active carbon/sulfur jacket immersed in a liquid ion electrolyte with a composite cathode. The research results suggest that sulfur undergoes a conversion reaction in the electrolyte in the solid body. The analysis of kinetics indicates that the slow, sulfur-stable reaction of the sulfur conversion results in a high voltage hysteresis and reduces the energy efficiency of the system.

According to Ref. [58], a battery using an aluminum-based redox combination, which includes three electron transfers in charge/discharge

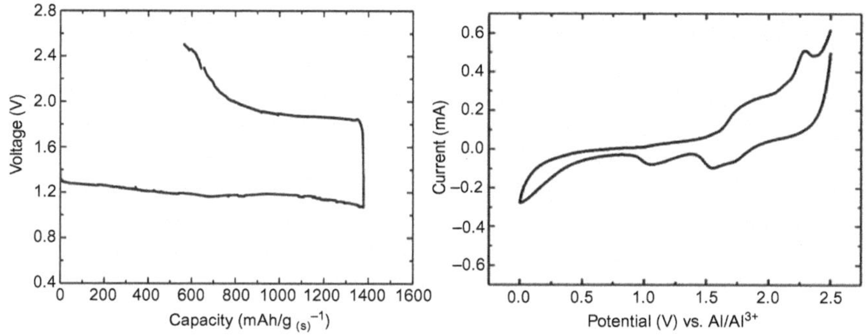

Fig. 50 Al-S battery characteristics [58].

electrochemical reactions, is desirable and attractive for research because of the low cost, the corresponding redox potential (E^0 (Al^{3+}/Al) = − 1.676 V NHE) and high charge density. The study described uses a new class of high-energy system, an aluminum-sulfur anhydrous battery. This battery included an anode from metallic Al and a mixture of sulfur and carbon as a cathode material. The system used as an electrolyte a mixture of $AlCl_3$ and ionic liquid EMI · Cl, where its composition controls its Lewis acidity. Preliminary results showed that the Al-S cell (Fig. 50) has a discharging voltage of about 1.2 V, and the specific energy was greater than 1700 Wh kg^{-1}, exceeding that found in most of the currently developed batteries. The battery operation mechanism, product properties, and limiting factors were investigated. This type of battery can be cheaper compared to current Li-ion and Li-S batteries and less susceptible to safety problems.

Recently Ref. [59] has presented a highly reversible room-temperature Al-S battery with a lithium-ion (Li^+-ion)-mediated ionic liquid electrolyte. Mechanistic studies with electrochemical and spectroscopic methodologies revealed that the enhancement in reversibility by Li^+-ion mediation is attributed to the chemical reactivation of aluminum polysulfides and/or sulfide by Li^+ during electrochemical cycling. The presence of a Li_3AlS_3-like product with a mixture of Li_2S- and Al_2S_3-like phases in the discharged sulfur cathode was found. With Li^+-ion mediation, the cycle life of room-temperature Al-S batteries was significantly improved. The cell delivered an initial capacity of ~1000 mAh g^{-1} and maintained a capacity of up to 600 mAh g^{-1} after 50 cycles.

REFERENCES

[1] G. Pistoia, Battery Operated Devices and Systems—From Portable Electronics to Industrial Products, Elsevier, 2009. Online version available at:http://app.knovel.com/hotlink/toc/id:kpBODSFPE1/battery-operated-devices/battery-operated-devices.

[2] G.L. Henricksen, A.N. Jansen, in: D. Linden, T.B. Reddy (Eds.), Handbook of Batteries, McGraw-Hill, New York, 2002 (Chapter 1).
[3] P.C. Symons, P.C. Butler, in: D. Linden, T.B. Reddy (Eds.), Handbook of Batteries, McGraw-Hill, New York, 2002 (Chapter 37).
[4] J.W. Braithwaite, W.L. Auxer, in: D. Linden, T.B. Reddy (Eds.), Handbook of Batteries, McGraw-Hill, New York, 2002 (Chapter 40).
[5] V. Dorval, C. St-Pierre, A. Valle´e, Lithium-Metal-Polymer Batteries: From the Electro-chemical Cell to the Integrated Energy Storage System (Avestor Report), 2004.
[6] C. St-Pierre, T. Gauthier, M. Hamel, M. Leclair, M. Parent, M.S. Davis, Avestor Lithium-Metal-Polymer Batteries Proven Reliability Based on Customer Field Trials (Avestor Report), 2003.
[7] V.S. Bagotsky, A.M. Skundin, Y.M. Volfkovich, Electrochemical Power Sources: Batteries, Fuel Cells and Supercapacitors. Part 15. Batteries with Molten Salt Electrolytes, first ed., John Wiley & Sons, Inc., 2013 (Published 2015 by John WiIey & Sons, Inc.).
[8] A.G. Ritchie, P.G. Bowles, D.P. Scattergood, Lithium-ion/iron sulphide rechargeable batteries, J. Power Sources 136 (2) (2004) 276–280.
[9] C.D.S. Tuck, Modern Battery Technology, Ellis Horwood, 1991.
[10] H.V. Venkatasetty, Lithium Battery Technology, John Wiley, 1984.
[11] K. Hansen, K. West, Lithium insertion into iron sulphides, Electrochem. Soc. Proc. 97-18 (1997) 124–132.
[12] R.A. Sharma, Equilibrium phases between lithium sulphide and iron sulphides, J. Electrochem. Soc. 123 (1976) 448–453.
[13] K. Takada, Y. Kitami, T. Inada, A. Kajiyama, M. Kouguchi, S. Kondo, M. Watanabe, M. Tabuchi, Electrochemical reduction of Li_2FeS_2 in solid electrolyte, J. Electrochem. Soc. 148 (2001)A1085.
[14] A.G. Ritchie, P. Bowles, Producing lithiated transition metal sulphides, UK Patent Application GB2351075, International Patent Application WO 00/78673.
[15] A.G. Ritchie, P. Bowles, Synthesis of lithium transition metal sulphides, UK Patent Application GB0029958.6, International Patent Application WO 02/46012.
[16] D.A.J. Rand, R. Woods, R.M. Dell, Batteries for Electric Vehicles, Research Study Press, Taunton, UK, 1998.
[17] S.M. Oweis, J. Embrey, J. Willson, P. Alunans, Rechargeable lithium-aluminum/iron sulfide batteries for high energy density, in: IEEE 35th International Power Sources Symposium, 1992.
[18] R.O. Ivins, E.C. Gray, W.J. Walsh, A.A. Chilenskas, Design, and performance of Li-Al/FeS batteries, in: Proceedings of 27th Power Sources Symposium, 1976, p. 8.
[19] F.J. Martino, T.D. Kaun, H. Shimotake, E.C. Gay, Advances in the development of lithium-aluminium-metalsulfide batteries for electric vehicle batteries, in: Proceedings of 13th Intersociety Energy Conversion Engineering Conference, 1978, p. 709.
[20] Z. Johnson, D. Pickett, S. Preston, B. Burns, R. Guidotti, High Temperature Battery for Space Applications, PSC-1 (OSPSC-21).
[21] R.A. Guidotti, G. Scharrer, F.W. Reinhard, Development of a high-power and high-energy thermal battery, in: Proceedings of the 39th Power Sources Conference, Cherry Hill, NJ, 12–15 June, 2000, pp. 547–551.
[22] H. Huang, J. Gao, L. Zhang, H. Hu, K. Zhu, A study of the CoS_2 cathode for thermal batteries, ECS Trans. 35 (32) (2011) 295–299.
[23] R.A. Guidotti, F.W. Reinhardt, J.X. Dai, D.E. Reisner, Preparation and characterization of nanostructured FeS_2 and CoS_2 for high-temperature batteries, Mater. Res. Soc. Symp. Proc. 730 (2002) 731–736.
[24] S. Xie, Y. Deng, J. Mei, Z. Yang, W.-M. Lau, H. Liu, Carbon coated CoS_2 thermal battery electrode material with enhanced discharge performances and air stability, Electrochim. Acta 231 (2017) 287–293.

[25] Y. Xie, Z. Liu, H. Ning, H. Huang, L. Chen, Suppressing self-discharge of Li–B/CoS_2 thermal batteries by using a carbon-coated CoS_2 cathode, RSC Adv. 8 (2018) 7173.
[26] J.L. Payne, J.D. Percival, K. Giagloglou, C.J. Crouch, G.M. Carins, R.I. Smith, R. Comrie, R.K.B. Gover, J.T.S. Irvine, In-situ thermal battery discharge using NiS_2 as a cathode material, ChemElectroChem 4 (2017) 1916–1923.
[27] P.J. Masset, R.A. Guidotti, Thermal activated ("thermal") battery technology: part IIIb. Sulfur and oxide-based cathode materials, J. Power Sources 178 (2008) 456–466.
[28] L. Yu, H. Hu, H.B. Wu, X.W. Lou, Complex hollow nanostructures: synthesis and energy-related applications, Adv. Mater. 29 (2017) 1604563–1604602.
[29] X. Xu, W. Liu, Y. Kim, J. Cho, Nanostructured transition metal sulfides for lithium ion batteries: progress and challenges, Nano Today 9 (2014) 604–630.
[30] X. Liu, J.Q. Huang, Q. Zhang, L. Mai, Nanostructured metal oxides and sulfides for lithium-sulfur batteries, Adv. Mater. (2017) 1601759.
[31] X.-Y. Yu, L. Yu, X.W.D. Lou, Metal sulfide hollow nanostructures for electrochemical energy storage, Adv. Energy Mater. 6 (2016) 1501333–1501347.
[32] X. Li, K. Qian, Y. He, C. Liu, D. An, Y. Li, D. Zhou, B. Li, Q.H. Yang, F. Kang, A dual-functional gel-polymer electrolyte for lithium ion batteries with superior rate and safety performances, J. Mater. Chem. A 5 (2017) 18888–18895.
[33] B. Jiang, Y. He, B. Li, S. Zhao, S. Wang, Y.B. He, Z. Lin, Polymer-templated formation of polydopamine-coated SnO_2 nanocrystals: anodes for cyclable lithium-ion batteries, Angew. Chem. Int. Ed. 56 (2017) 1869–1972.
[34] Y. Li, S. Wang, Y.B. He, L. Tang, Y.V. Kaneti, W. Lv, Z. Lin, B. Li, Q.H. Yang, F. Kang, Li-ion and Na-ion transportation and storage properties in various sized TiO_2 spheres with hierarchical pores and high tap density, J. Mater. Chem. A 5 (2016) 4359–4367.
[35] C. Jin, L. Zhou, L. Fu, J. Zhu, D. Li, W. Yang, The acceleration intermediate phase (NiS and Ni_3S_2) evolution by nanocrystallization in Li/NiS_2 thermal batteries with high specific capacity, J. Power Sources 352 (2017) 83–89.
[36] S. Xie, Y. Deng, J. Mei, Z. Yang, W.-M. Lau, H. Liu, Facile synthesis of CoS_2/CNTs composite and its exploitation in thermal battery fabrication, Compos. Part B 93 (2016) 203–209.
[37] Y. Ji, X. Liu, W. Liu, Y. Wang, H. Zhang, M. Yang, X. Wang, X. Zhao, S. Feng, A facile template-free approach for the solid-phase synthesis of CoS_2 nanocrystals and their enhanced storage energy in supercapacitors, RSC Adv. 4 (2014) 50220–50225.
[38] Z. Yang, X. Liu, X. Feng, Y. Cui, X. Yang, Hydrothermal synthesized micro/nano-sized pyrite used as cathode material to improve the electrochemical performance of thermal battery, J. Appl. Electrochem. 44 (2014) 1075–1080.
[39] C. Jin, L. Fu, J. Zhu, W. Yang, D. Li, L. Zhou, A hierarchical carbon modified nano-NiS_2 cathode with high thermal stability for a high energy thermal battery, J. Mater. Chem. A (2018), https://doi.org/10.1039/C8TA00346G (Paper; Advance Article).
[40] F.H. Zahrul, W.C. Lee, F.M.S. Mohd, B.I. Amir, Modeling of sodium sulfur battery for power system applications, Elektrika 9 (2) (2007) 66–72.
[41] M. Kamibayashi, K. Tanaka, Recent sodium sulfur battery applications, in: Proc. IEEE PES Transmission and Distribution Conference and Exposition, USA, 28 October–2 November, vol. 2, 2001, pp. 1169–1173.
[42] B. Tamyurek, D.K. Nichols, O. Demirci, The NAS battery: a multi-function energy storage system, in: Proc. IEEE PES General Meeting, USA, 13–17 July, vol. 4, 2003.
[43] F.M. Stackpool, W. Auxer, M. McNamee, M.F. Mangan, Sodium sulphur battery development, in: Proc. 24th Intersociety Energy Conversion Engineering Conference, IECEC-89, August 6–11, vol. 6, 1989, pp. 2765–2768.
[44] K. Takashima, F. Ishimaru, A. Kunimoto, H. Kagawa, K. Matsui, E. Nomura, Y. Matsumaru, A. Kita, S. Iijima, T. Kato, Y. Mat-Suo, T. Nakayama, Y. Sera, A plan for a 1MW/8MWH sodium-sulfur battery energy storage plant, in: Proc. 25th Inter-

Society Energy Conversion Engineering Conference, IECEC-90, August 12–17, vol. 3, 1990, pp. 367–371.
[45] H. Takami, B.I. Amir, F.M.S. Mohd, NAS battery energy storage system for power quality support in Malaysia, in: Proc. International Power & Energy Conference, IPEC2003 Conf., November 28, 2003.
[46] D. Linden, T. Reddy, Handbook of Batteries, McGraw-Hill, third ed., 2002, pp. 40.1–40.23.
[47] J. Liang, F. Li, H.-M. Cheng, Batteries with a sulfur cathode: a leap forward in energy density, Energy Storage Mater. 8 (2017) A1–A3.
[48] I.-S. Ahn, X. Liu, H.-J. Ahn, Synthesis of cathode material-nickel sulfides by mechanical alloying for sodium batteries, J. Korean Powder Metall. Inst. 19 (3) (2012) 182–188.
[49] Q. Lu, X. Wang, J. Cao, C. Chen, K. Chen, Z. Zhao, Z. Niu, J. Chen, Freestanding carbon fiber cloth/sulfur composites for flexible room-temperature sodium-sulfur batteries, Energy Storage Mater. 8 (2017) 77–84.
[50] L. Zeng, Y. Yao, J. Shi, Y. Jiang, W. Li, L. Gu, Y. Yu, A flexible $S_{1-x}Se_x$@porous carbon nanofibers ($x \leq 0.1$) thin film with high performance for Li-S batteries and room-temperature Na-S batteries, Energy Storage Mater. 5 (2016) 50–57.
[51] S. Wei, S. Xu, A. Agrawral, S. Choudhury, Y. Lu, Z. Tu, L. Ma, L.A. Archer, A stable room-temperature sodium–sulfur battery, Nat. Commun. 7 (2016), 11722 https://doi.org/10.1038/ncomms11722.
[52] K. Siczek, K. Siczek, Evaluation of the possibility of a sodium-sulfur battery cooperating with an internal combustion engine, Buses, Exploitation and tests 6/2017.
[53] S. Licht, D. Peramunage, Novel aqueous aluminum/sulfur batteries, J. Electrochem. Soc. 140 (1) (1993) L4–L6.
[54] D. Peramunage, S. Licht, A solid sulfur cathode for aqueous batteries, Science 261 (5124) (1993) 1029–1032.
[55] G. Levitin, C. Yarnitzky, S. Licht, Fluorinated graphites as energetic cathodes for non-aqueous Al batteries, Electrochem. Solid-State Lett. 5 (7) (2002) A160–A163.
[56] G. Cohn, L. Ma, L.A. Archer, A novel non-aqueous aluminum sulfur battery, J. Power Sources 283 (2015) 416–422.
[57] T. Gao, X. Li, X. Wang, J. Hu, F. Han, X. Fan, L. Suo, A.J. Pearse, S.B. Lee, G.W. Rubloff, K.J. Gaskell, M. Noked, C. Wang, A rechargeable Al/S battery with an ionic-liquid electrolyte, Angew. Chem. Int. Ed. 55 (2016) 9898–9901.
[58] G. Cohn, L.A. Archer, High energy non-aqueous Al-S battery, in: ECS and SMEQ Joint International Meeting, October 5–9, Canun, Mexico, 2014.
[59] X. Yu, M.J. Boyer, G.S. Hwang, A. Manthiram, Room-temperature aluminum-sulfur batteries with a lithium-ion-mediated ionic liquid electrolyte, Chem 4 (3) (2018) 586–598.

FURTHER READING

[60] G. Pistoia, Nonaqueous batteries used in industrial applications, in: M. Broussely, G. Pistoia (Eds.), Industrial Applications of Batteries. From Cars to Aerospace and Energy Storage, Elsevier, Amsterdam, 2007 (Chapter 1).
[61] A. Gilmour, C.O. Giwa, J.C. Lee, A.G. Ritchie, Lithium rechargeable envelope cells, in: Power Sources 16, Proceedings of the 20th International Power Sources Symposium, Brighton, April, J. Power Sources 65 (1997) 219–224.

CHAPTER TWELVE

Mg-S Batteries

According to Ref. [1], the metal magnesium (Mg) is a promising material for anodes in next-generation batteries, due to its nondendritic deposition and high capacity. However, the best cathode for the Mg accumulator was based on the $Mg_xMo_3S_4$ complex, with high molecular weight, which made the full link energetically uncompetitive. To increase the energy density, a high-capacity cathode material in the form of sulfur was proposed. However, until now only limited research has been done on the Mg/S system, and all of the studies are plagued by the weak reversibility attributed to the formation of electrochemically inactive forms of MgS_x. A new strategy was developed, based on the use of the Li^+ role in activating the MgS_x form, for the combination of a nondendritic Mg anode with a reversible polysulfide cathode, and a truly reversible Mg-S battery with a capacity of up to $1000\,mAh\,g^{-1}$ for more than 30 cycles. The cell showed two discharge plates at 1.75 and 1.0 V, which corresponds to an energy density of up to $874\,Wh\,kg^{-1}$. It was shown that the Mg/S battery has a much better reversibility after the introduction of LiTFSI as an additive. It has been suggested that the reversibility results from chemical reactivation of MgS_x (more specifically MgS and MgS_2) with Li^+.

The diagram of Mg-S cell operation is shown in Fig. 51. Fig. 52A shows charging/discharging curves for a sulfur cathode in an electrolyte with 0.1 M Mg-HMDS + 1.0 M LiTFSI composition in three electrode cells with a current of $71\,mAh\,g^{-1}$ at room temperature. Fig. 52B illustrates the stability of the cyclic Mg-S battery operation in the electrolyte with and without LiTFSI [1].

The mechanism of operation of the Mg-S battery with the addition of LiTFSI is shown in Fig. 53.

According to Ref. [2], magnesium accumulators (Mg) have been proposed for electrochemical storage of energy, and for use in electric vehicles, due to the properties of Mg, such as prevalence of occurrence, work safety, and high volume capacity. Among the various anode materials of alkaline/alkaline earth metals, magnesium (Mg) has the highest theoretical volumetric capacity ($3832\,mAh\,cm^{-3}$) and a high negative reduction potential of $-2.356\,V$ compared to a normal hydrogen electrode (NHE).

Next-generation Batteries with Sulfur Cathodes
https://doi.org/10.1016/B978-0-12-816392-4.00012-8

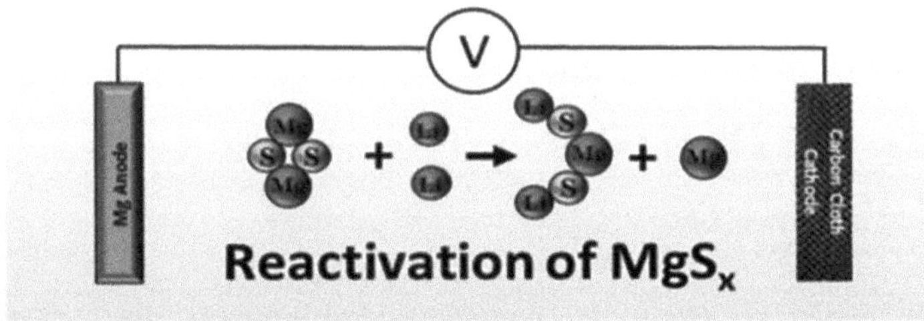

Fig. 51 Diagram of the Mg-S cell [1].

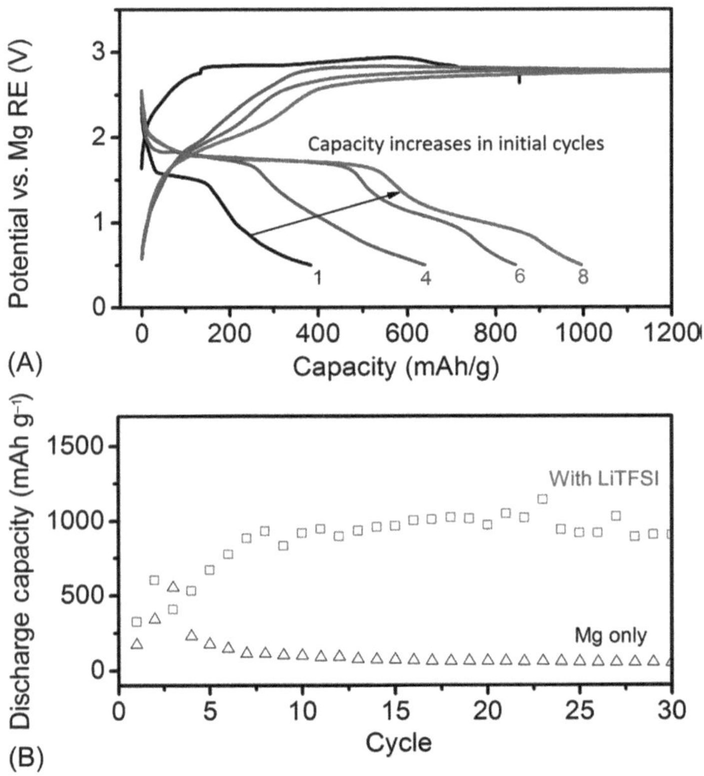

Fig. 52 (A) Load/discharge curves for a sulfur cathode in an electrolyte composition 0.1 M Mg-HMDS + 1.0 M LiTFSI in three electrode cells for a current of 71 mAh g^{-1} at room temperature. The *arrow* shows the increasing capacity tendency for the ACC/S composite cathode because of slow electrolyte penetration. (B) Stability of the cyclic Mg-S battery in electrolyte with and without LiTFSI [1].

For comparison, graphite anodes currently used in lithium-ion batteries (LIB) have a volumetric efficiency of only 777 mAh cm^{-3} [3]. As an anode material, Mg can be considered a safer electrode in liquid electrolytes,

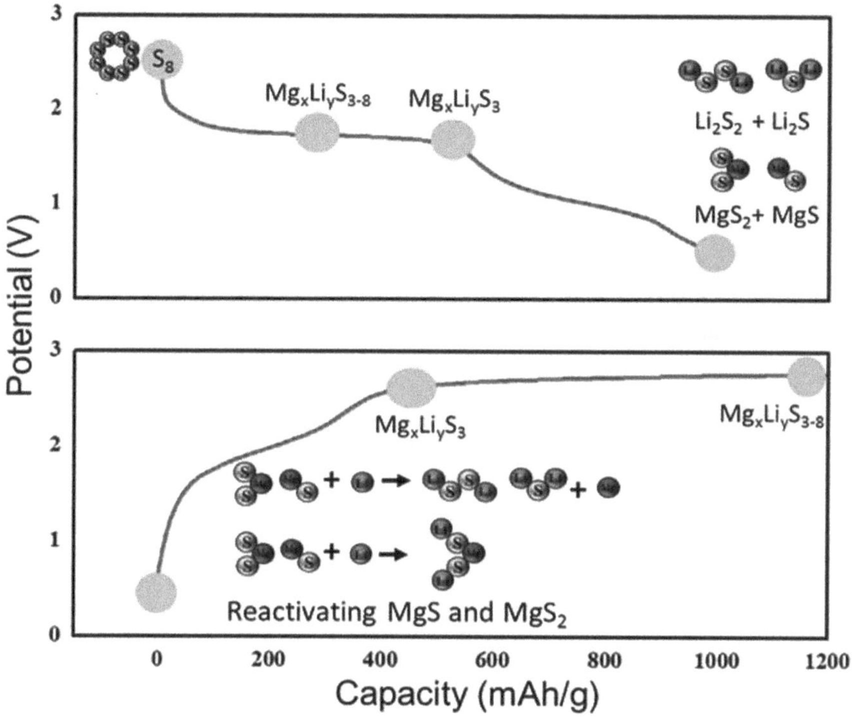

Fig. 53 Mechanism of Mg/S battery operation with the addition of LiTFSI [1].

because it does not form dendrites as compared to lithium [4]. Theoretical calculations show that Mg promotes the growth of smooth surfaces compared to Li and Na, due to its lower diffusion barriers and higher-coordinated configurations [5]. Raw material for Mg is cheaper than for lithium, and its compounds are usually nontoxic.

Sulfur (S) is a promising cathode battery material based on Mg, due to the high theoretical capacity ($1671\,mAh\,g^{-1}$ or $3459\,mAh\,cm^{-3}$) and natural prevalence. The combination of the Mg anode and the sulfur cathode gives a theoretical energy density of $3200\,Wh\,L^{-1}$, compared to $2800\,Wh\,L^{-1}$ for lithium-sulfur batteries [3,6].

However, several problems must be solved for the Mg/S battery. The limits of development of Mg/S batteries result from the limited availability of suitable electrolytes with high ionic conductivity, in which Mg can be reassembled [7]. This is mainly due to the strong electrophilic nature of sulfur, which requires a nonnucleophilic electrolyte. We developed a non-nucleophilic electrolyte by adding $AlCl_3$ to the THF solution of magnesium hexamethyldisilazide chloride and tested the Mg/S cell in two cycles with a discharging capacity of 1200 and 394 mAh g^{-1}, respectively, at a discharging voltage of about 1 V [3]. Recently, information about a nonnucleophilic Mg/S battery electrolyte based on a nonnucleophilic magnesium-bis(hexamethyldisilazide) [(HMDS) ^{2}Mg]-reacted with $AICI_3$ as Lewis acid (1:2 molar ratio) was reported, and the reaction products were dissolved in other solvents, such as ether compounds [8]. Using these electrolytes, the potential of discharging the Mg/S battery was increased to 1.65 V, which is close to the thermodynamic voltage (~1.7 V) [9].

According to Ref. [10] most simple magnesium salts tend to passivate the Mg metal surface too quickly to function as electrolytes for Mg batteries. To improve such situations, an electroactive salt $[Mg(THF)_6][AlCl_4]_2$ was synthesized and structurally characterized. The Mg electrolyte based on this simple mononuclear salt showed a high Mg cycling efficiency, good anodic stability (2.5 V vs. Mg), and high ionic conductivity (8.5 mS cm^{-1}). Magnesium/sulfur cells employing the as-prepared electrolyte exhibited good cycling performance over 20 cycles in the range of 0.3–2.6 V, thus indicating an electrochemically reversible conversion of S to MgS without severe passivation of the Mg metal electrode surface.

According to Ref. [11], room temperature rechargeable magnesium (Mg) batteries were designed from Mg as a negative material, sulfur-containing composite prepared from elemental sulfur and the bis(alkenyl) compound having a crown ether unit ($BUMB_{18}C_6$) or linear ether unit (UOEE) as a positive material and the simple electrolyte (0.7 mol dm^{-3} $Mg[N(SO_2CF_3)_2]_2$-triglyme (G_3) solution). The reaction between molten S and the bis(alkenyl) compound ($BUMB_{18}C_6$ or UOEE) provided the sulfur-containing composite, S-BUMB18C6 or S-UOEE. Both sulfur-containing composites were electrochemically active in the Mg salt-based electrolyte, acetonitrile- or G_3-$Mg[N(SO_2CF_3)_2]_2$ electrolyte. The first discharge capacity of the test cells with the sulfur-containing composite was 460 Ah kg^{-1} with an S-$BUMB_{18}C_6$ electrode and 495 Ah kg^{-1} with the S-UOEE electrode. According to the continuous charge-discharge cycle tests (at the 10th cycle), the discharge capacity of the test cell with the S-$BUMB_{18}C_6$ electrode (68.1 Ah kg^{-1}) was higher than that with the S-UOEE electrode (0.18 Ah kg^{-1}). The crown ether units in the S-$BUMB_{18}C_6$ composite created ion-conducting paths in the cathode, preventing a rise in the internal resistance of the cathode, and providing better cycle performance with the test cells having the S-$BUMB_{18}C_6$ composite electrode than those with the S-UOEE electrode.

The first tests of the Mg/S battery pointed to the effects of polarization during charging, low stability of cyclic operation, disappearance of the initial capacity, and dissolution of polysulfides, analogous to the Li/S battery system [12,13]. In Li/S batteries, soluble polysulfides led to self-discharge and degradation of the cell [14,15]. The electrical insulation property of sulfur is another problem that gives low electronic transmission of the load to the current collectors. To solve some of these problems, it is necessary to design a nanodispersed active sulfur composite in a suitable carbon matrix host. The carbonaceous host material must have (a) a strong chemical anchoring of sulfur and then form polysulfides, (b) a high electrical conductivity, (c) a mechanically stable frame withstanding the load generated by sulfur volume changes during cyclic operation, (d) easy access of liquid electrolyte to active sulfur, and (e) small pores without large holes to contain polysulfides. Among the various types of carbon nanomaterials there are graphene-based nanostructures (G), which are preferred materials for energy storage applications, due to their unique properties, such as large surface area, high load mobility, excellent electronic and thermal conductivity, high mechanical strength, and good stability chemically [16–18].

According to Ref. [2], the magnesium Mg battery was produced using a graphene-sulfur nanocomposite as a cathode, a magnesium-carbon composite as an anode, and as an electrolyte a nonnucleophilic Mg-based complex in a tetraglyme solvent. The graphene-sulfur nanocomposite was prepared by combining thermal and chemical precipitation methods. The Mg-S cell provides a higher reversible capacity (448 $mAh\,g^{-1}$), a longer cyclic capacity (236 $mAh\,g^{-1}$ at the end of the 50th cycle), and a better ability to withstand rated currents than previous cells. The use of a cathode from a graphene-sulfur composite with properties such as a large surface, porous morphology, very good electronic conductivity, the presence of functional oxygen groups, together with a nonnucleophilic magnesium electrolyte ensures greater efficiency of the battery.

Fig. 54A shows the stability of the cyclic S-rGO nanocomposite for the first 50 cycles at a current density of 20 $mA\,g^{-1}$. In turn, Fig. 54B shows the ability to withstand the nominal currents of the S-rGO electrode for current densities between 5 and 45 $mA\,g^{-1}$. Fig. 55 shows the electrochemical mechanism of the nanocomposite S-rGO electrode inside the Mg/S cell.

During the study described in Ref. [19], vanadium oxychloride was used as an electrode material in a Mg-based battery. The cell delivered just 45 $mA\,h\,g^{-1}$ in the first cycle. The delivered capacity could be improved

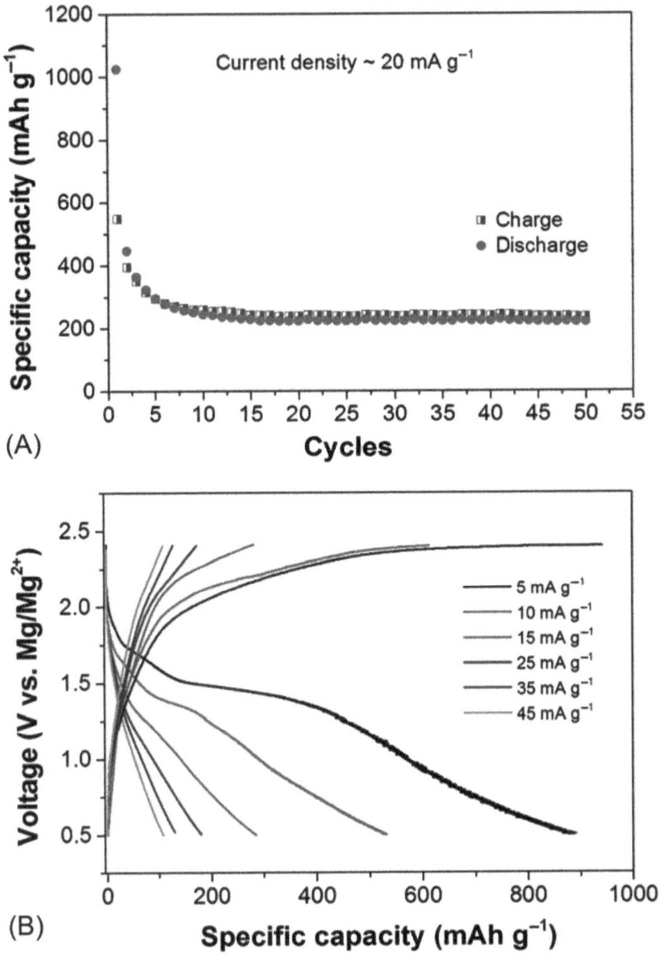

Fig. 54 (A) Stability of the cyclic S-rGO nanocomposite for the first 50 cycles at a current density of 20 mA g^{-1}. (B) Ability to withstand the nominal currents of the S-rGO electrode for current density between 5 and 45 mA g^{-1} [2].

through preliminary cycling of the VOCl electrode with Li, as the VOCl can expand its interlayer spacing upon intercalation of ions or molecules within them. In fact, a VOCl electrode with expanded interlayer spacing should facilitate the intercalation of Mg^{2+}, thus leading to higher specific capacities. The Li pretreatment was able to promote the specific capacity by a factor of four (170 mAh g^{-1}) after the first discharge at 298 K. Over 130 mAh g^{-1} was retained at 5 mA g^{-1} after 70 cycles.

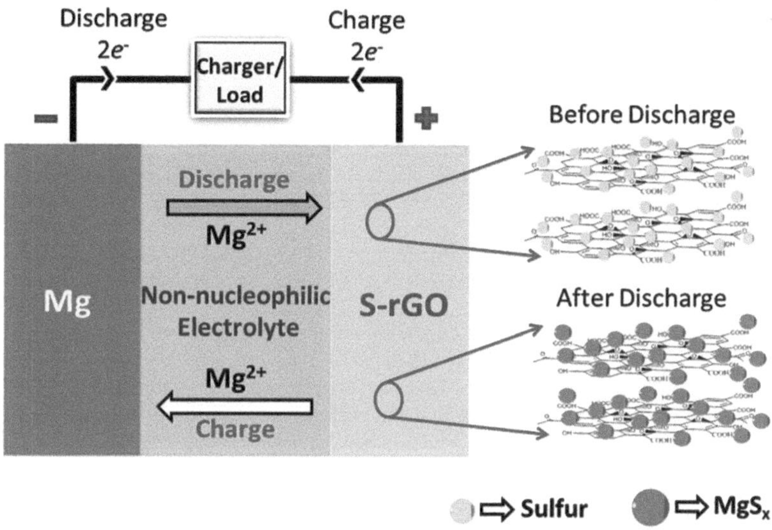

Fig. 55 Electrochemical mechanism of the nanocomposite S-rGO electrode inside the Mg/S cell [2].

REFERENCES

[1] T. Gao, M. Noked, A.J. Pearse, E. Gillette, X. Fan, Y. Zhu, C. Luo, L. Suo, M.A. Schroeder, K. Xu, S.B. Lee, G.W. Rubloff, C. Wang, Enhancing the reversibility of Mg/S battery chemistry through Li^+ mediation. J. Am. Chem. Soc. 137 (38) (2015) 12388–12393, https://doi.org/10.1021/jacs.5b07820.

[2] B.P. Vinayan, Z. Zhao-Karger, T. Diemant, V.S.K. Chakravadhanula, N.I. Schwarzburger, M.A. Cambaz, R.J. Behm, C. Kübel, M. Fichtner, Performance study of magnesium–sulfur battery using a graphene based sulfur composite cathode electrode and a non-nucleophilic Mg electrolyte, Nanoscale 8 (2016) 3296–3306, https://doi.org/10.1039/C5NR04383B (Paper).

[3] H.S. Kim, T.S. Arthur, G.D. Allred, J. Zajicek, J.G. Newman, A.E. Rodnyansky, A.G. Oliver, W.C. Boggess, J. Muldoon, Structure and compatibility of a magnesium electrolyte with a sulphur cathode, Nat. Commun. 2 (2011) 427.

[4] T.D. Gregory, R.J. Hoffman, R.C. Winterton, Nonaqueous electrochemistry of magnesium applications to energy storage, J. Electrochem. Soc. 137 (1990) 775–780.

[5] M. Jäckle, A. Groß, Microscopic properties of lithium, sodium, and magnesium battery anode materials related to possible dendrite growth, J. Chem. Phys. 141 (2014)174710.

[6] X. Ji, K.T. Lee, L.F. Nazar, A highly ordered nanostructured carbon–sulphur cathode for lithium–sulphur batteries, Nat. Mater. 8 (2009) 500–506.

[7] O. Tutusaus, R. Mohtadi, Paving the way towards highly stable and practical electrolytes for rechargeable magnesium batteries, ChemElectroChem 2 (2015) 51–57.

[8] Z. Zhao-Karger, X. Zhao, O. Fuhr, M. Fichtner, Bisamide based non-nucleophilic electrolytes for rechargeable magnesium batteries, RSC Adv. 3 (2013) 16330–16335.

[9] Z. Zhao-Karger, X. Zhao, D. Wang, T. Diemant, R.J. Behm, M. Fichtner, Performance improvement of magnesium sulfur batteries with modified non-nucleophilic electrolytes, Adv. Energy Mater. 5 (2014)1401155.

[10] W. Li, S. Cheng, J. Wang, Y. Qiu, Z. Zheng, H. Lin, S. Nanda, Q. Ma, Y. Xu, F. Ye, Synthesis, crystal structure, and electrochemical properties of a simple magnesium electrolyte for magnesium/sulfur batteries, Angew. Chem. Int. Ed. 55 (2016) 6406–6410, https://doi.org/10.1002/anie.201600256.
[11] K. Itaoka, I.T. Kim, K. Yamabuki, N. Yoshimoto, H. Tsutsumi, Room temperature rechargeable magnesium batteries with sulfur-containing composite cathodes prepared from elemental sulfur and bis(alkenyl) compound having a cyclic or linear ether unit. J. Power Sources 297 (2015) 323–328, https://doi.org/10.1016/j.jpowsour.2015.08.029.
[12] P. Saha, M.K. Datta, O.I. Velikokhatnyi, A. Manivannan, D. Alman, P.N. Kumta, Rechargeable magnesium battery: current status and key challenges for the future, Prog. Mater. Sci. 66 (2014) 1–86.
[13] A. Manthiram, Y. Fu, Y.-S. Su, Challenges and prospects of lithium–sulfur batteries, Acc. Chem. Res. 46 (2012) 1125–1134.
[14] D. Moy, A. Manivannan, S.R. Narayanan, Direct measurement of polysulfide shuttle current: a window into understanding the performance of lithium-sulfur cells, J. Electrochem. Soc. 162 (2015) A1–A7.
[15] Y.V. Mikhaylik, J.R. Akridge, Polysulfide shuttle study in the Li/S battery system, J. Electrochem. Soc. 151 (2004) A1969–A1976.
[16] Y. Sun, Q. Wu, G. Shi, Graphene based new energy materials, Energy Environ. Sci. 4 (2011) 1113–1132.
[17] J. Hou, Y. Shao, M.W. Ellis, R.B. Moore, B. Yi, Graphene-based electrochemical energy conversion and storage: fuel cells, supercapacitors and lithium ion batteries, Phys. Chem. Chem. Phys. 13 (2011) 15384–15402.
[18] B.P. Vinayan, N.I. Schwarzburger, M. Fichtner, Synthesis of a nitrogen rich (2D–1D) hybrid carbon nanomaterial using a MnO_2 nanorod template for high performance Li-ion battery applications, J. Mater. Chem. A 3 (2015) 6810–6818.
[19] C.B. Minella, P. Gao, Z. Zhao-Karger, X. Mu, T. Diemant, M. Pfeifer, V.S.K. Chakravadhanula, R.J. Behm, M. Fichtner, Interlayer-expanded vanadium oxychloride as an electrode material for magnesium-based batteries, ChemElectroChem 4 (2017) 738.

CHAPTER THIRTEEN

Zn-S Batteries

Early in Ref. [1] an electric storage cell containing a zinc anode and a sulfur cathode was described, wherein the zinc and the sulfur were in contact with an aqueous solution containing sulfur during the process of battery discharge. Specific conditions for the aqueous sulfur electrolyte were chosen to overcome the normal ineffectiveness of zinc oxidation in the presence of aqueous zero valent sulfur. Normally, a zinc anode cannot be oxidized in an aqueous solution containing sulfur, because the product of the discharge would be zinc sulfide, which is a highly insoluble salt and creates a layer that passivates the zinc and renders it completely ineffective to battery discharge. The performance of the battery is made possible by high OH^- and HS-ion concentrations formed by the addition of salts to the aqueous zero valent sulfur solution and permits effective and efficient battery discharge.

In Ref. [2] an electrochemical cell with a common zinc anode as a negative pole and a sulfur cathode as a positive pole was studied, wherein the solid sulfur is located on an iron (II) sulfide electrode in an ammonium chloride solution. For better electron transfer to the suspended sulfur, fine powdered ferrous sulfide could be added to the sulfur suspension or the sulfur with the finely powdered iron (II) sulfide could be previously solid fused with the iron (II) sulfide electrode. Instead of a solid iron (II) sulfide electrode, a carbon electrode could also be used as the discharge electrode. However, it would then be necessary to use a mixture of powdered ferrous sulfide and sulfur as the active cathode material. The zinc anode and sulfur/iron (II) sulfide cathode were separated by a porous diaphragm.

It was shown that elemental sulfur at room temperature can serve as a cathode material in zinc-sulfur batteries, as it is rapidly reduced to iron (II) sulfide electrodes in aqueous electrolytes such as ammonium chloride or sodium chloride solutions. The quiescent voltage of a zinc-sulfur battery is 0.6V. Capacities and energy densities exceed those of zinc-brownstone cells.

According to Ref. [3], a new Zn/S water accumulator system consists of a zinc anode and a sulfur/polysulfide cathode located in a highly

Next-generation Batteries with Sulfur Cathodes
https://doi.org/10.1016/B978-0-12-816392-4.00013-X

concentrated alkaline electrolyte. High energy parameters (Faraday's theoretical capacity is 550 Ah kg^{-1} and theoretical energy density is 572 Wh kg^{-1}), cheapness, and low toxicity make this system competitive in relation to other existing water battery systems.

In aqueous polysulfurates, zinc anodes are completely passivated, due to the formation of an insulating ZnS semiconductor on the surface of the electrodes in the oxidation process, blocking the surface of the electrode and preventing further electrochemical reaction. In the presence of concentrated alkaline solutions (more than 6 M KOH), the surface of the zinc can be reactivated, and the zinc becomes an effective anode in the polysulfide environment.

A new functional zinc/polysulfide battery was built in the configuration of a two-chamber cell, the zinc anode chamber being separated from the polysulfide cathode chamber through a cation-permeable membrane. Due to its energy parameters, this configuration cannot compete with existing commercial and noncommercial battery systems. Therefore further research was directed toward zinc-sulfur cells with a single-chamber configuration.

Many different experiments were carried out using a "thick" single-chamber cell configuration. The best result (105 mA cm^{-2}) was obtained for a solution of 3 M K_2S_3 + 14 M KOH discharged under a load of 3.3 Ω and 50°C. This practical capacity of Faraday is competitive in comparison to those obtained for commercial primary water battery systems. On the other hand, it is only 20% of Faraday's theoretical capacity of the Zn/S battery system (550 Ah kg^{-1}).

To further improve Faraday's capacity, a thin single-chamber link was built. This thin configuration (total battery thickness around 0.3 mm) causes an approximately twofold increase in discharging capacity (175 Ah kg^{-1}) compared to other, thicker cell configurations.

A button configuration, using a solid sulfur cathode with the addition of electrocatalytic CoS, allowed a reduction in the total battery weight and further improvement in the measured discharging capacity to 280 Ah kg^{-1}, or more than 200 Wh kg^{-1} measured energy density.

In Ref. [4] it was reported that the power of a new zinc-sulfur cell (with generalized $Zn + S \rightarrow ZnS$ reaction) was tested to obtain a high (>1000 Ah kg^{-1}) sulfur capacity (load). At 25°C, the theoretical energy density of the entire Zn/S system was high at 572 Wh kg^{-1}, at $E° = 1.04$ V. From 273 to 373 K, the cell thermodynamic potential was calculated for 29 possible solution phase reactions in the system, including various known coexisting water form polysulfides. In this area, zinc sulfide products are thermodynamically more favorable than zinc products, which may be more advantageous compared to Al/S batteries, where aluminate is the preferred discharge product. Initial experimental investigations of the zinc-sulfur electrochemical system were carried out for the cell being an anode/water zeolite polysulfide system/electrocatalytic cathode with CoS. The measured extended cathode capacity approaches the theoretical capacity of 2 Faradays corresponding to the oxidation limit for sulfur. However, the anode tends to passivate due to the discharge product in the form of zinc sulfide, which prevents permanent discharge. Passivation is overcome in potassium polysulfide solutions containing concentrated (>10 mol) KOH; this results in an effective, long-lasting capacity (load/mol) of 2 Faradays/mol for zinc oxidation and high loading capacity of the zinc-sulfur system.

REFERENCES

[1] S. Licht, Zinc Sulfur Battery, US6207324B1.
[2] H. Helms, J. Ristau, H. Kunz, W. Jansen, Low-temperature zinc-sulfur battery, Monatshefte fuer Chemie (Chem. Mon.) 131 (2) (2000) 165–174.
[3] T.A. Bendikov, Novel Aqueous Zinc-Sulfur Battery (Ph.D. Thesis), Technion—Israel Institute of Technology, 2002.
[4] T.A. Bendikov, C. Yarnitzky, S. Licht, Energetics of a zinc–sulfur fuel cell. J. Phys. Chem. B 106 (11) (2002) 2989–2995, https://doi.org/10.1021/jp0135454.

CHAPTER FOURTEEN

All-Solid-State Metal-Sulfur Batteries

During studies described in López-Aranguren et al. [1], a metal hydride was used as an anode in a complete all-solid-state battery with S/C as cathode and $LiBH_4$ as a solid electrolyte. The hydride was a nanocomposite made of 0.8 MgH_2 and 0.2 TiH_2 counterparts. The anode half-cell showed discharge of 1600 mAh g^{-1}, i.e., conversion rate >90%. The battery had a high reversible capacity of 910 mAh g^{-1} with discharge plateaus at 1.8 and 1.4 V. The capacity remained to 85% of the initial value over the 25 first charge/discharge cycles.

In Ref. [2] all-solid-state lithium-sulfur batteries based on nanoconfined $LiBH_4$ in mesoporous silica as solid electrolytes were presented. The nanoconfined $LiBH_4$ had fast ionic lithium conductivity at room temperature, 0.1 mS cm^{-1}, negligible electronic conductivity, and its cationic transport number ($t^+=0.96$), close to unity, demonstrated a purely cationic conductor. The electrolyte had an excellent stability against lithium metal. The batteries showed very good performance, delivering high capacities vs. sulfur mass, typically 1220 mAh g^{-1} after 40 cycles at moderate temperature (55°C), 0.03 C rates, and working voltage of 2 V.

In Ref. [3] a three-dimensional (3D) bilayer garnet solid-state electrolyte framework for advanced Li metal batteries was demonstrated. The dense layer was reduced in thickness to a few microns and still retained good mechanical stability, thereby enabling the safe use of Li metal anodes. The thick porous layer acted as a mechanical support for the thin dense layer, which served as a host for high loading of cathode materials and provided pathways for continuous ion transport. Results showed that the integrated sulfur cathode loading can reach >7 mg cm^{-2} while the proposed hybrid Li-S battery exhibited a high initial coulombic efficiency (>99.8%) and high average coulombic efficiency (>99%) during the subsequent cycles.

According to Ref. [4], the performance of an all-solid-state lithium-sulfur battery, comprised of a positive composite electrode with high P/S ratio (number of phosphorus atoms/number of sulfur atoms) and solid electrolyte, was studied at 25°C at several loading weights of the positive

Next-generation Batteries with Sulfur Cathodes
https://doi.org/10.1016/B978-0-12-816392-4.00014-1

electrode. The activation energy was nearly independent of the loading weight of the positive composite electrode. It was found that the influence of the sulfur reactivity was significantly larger than those of the ionic and electric conductivities. The positive composite electrode provided a capacity of above 1550 mAh g^{-1} (sulfur) at 1.3 mA cm^{-2} and above 6600 W kg^{-1} (sulfur) at 50% state-of-charge for a loading weight of 8.3 mg cm^{-2}. With this positive electrode, the energy and power densities were 520 Wh kg^{-1} (970 Wh L^{-1}) and 1230 W kg^{-1} (2300 W L^{-1}), respectively, at the cell level (18,650). The cycling performance of the cell with such composite positive electrode was high over several hundred cycles.

In Ref. [5] a positive composite electrode prepared using P_2S_5 and a solid electrolyte with a high ionic conductivity was studied. The lithium-sulfur cell exhibited a relatively low activation energy together with high ionic conductivity. The positive composite electrode exhibited an extremely high capacity of 1550 mAh g^{-1} (sulfur) at 1.3 mA cm^{-2} and 25°C.

The ionic conductivity of $Li_{1.5}PS_{3.3}$, a solid electrolyte (SE) with a high P/S ratio, was greatly improved by the addition of LiI. An SE with the formula $0.82(Li_{1.5}PS_{3.3})0.18(LiI)$ showed the highest ionic conductivity (0.31 mS cm^{-1}) of the compounds studied. An SE with a combination of P_2S_5 acting as an SE with a high P/S ratio and $0.82(Li_{1.5}PS_{3.3})0.18(LiI)$ acting as an SE with high ionic conductivity improved battery performance and both the sulfur reactivity and the ionic conductivity of the positive composite electrode. A cell prepared using this positive composite electrode showed high cycling performance over 100 cycles at a practical current density. The mass- and volume-based energy densities of a cell comprising a positive composite electrode prepared from $0.82(Li_{1.5}PS_{3.3})0.18(LiI)$ were 540 Wh kg^{-1} and 990 Wh L^{-1}, respectively, at the cell level. Such a positive composite electrode has sufficient potential for large-scale power storage equipment and electric vehicles.

According to Ref. [6], all-solid-state lithium-sulfur batteries using a composite electrode containing activated carbon with enhanced electroconductivity were studied at 25°C. The incorporation of poly(3,4-ethylenedioxythiophene)/poly(styrenesulfonate) into activated carbon enhanced conductivity, improving battery performance. The cell reversible capacity exceeded 1300 mAh g^{-1} at 0.64 mA cm^{-2} and 25°C, despite a high sulfur content (60 wt%). The obtained increase of electrical conductivity because of the conductive polymer material was higher during charging than during discharging.

As reported in Ref. [7], the all-solid-state lithium secondary battery using a $NiPS_3$ electrode and an $80Li_2S{\cdot}20P_2S_5$ glass solid electrolyte was elaborated. This all-solid-state cell exhibited a stable reversible capacity of about $80\,mAh\,g^{-1}$ after 30 cycles with a current density of $64\,\mu A\,cm^{-2}$ at room temperature. It was found that $NiPS_3$ can be an electrode active material in all-solid-state lithium secondary batteries using Li_2S-P_2S_5 solid electrolytes.

During studies described in Ref. [8], an all-solid-state Na-S battery at 25°C was used, in which the sulfur content in the positive composite electrode was 50 wt% to enhance energy density. All-solid-state Na-S cells prepared using a positive composite electrode with a high-phosphorus-to-sulfur ratio solid electrolyte exhibited a high capacity of approximately $1500\,mAh\,g^{-1}$ at $0.13\,mA\,cm^{-2}$ at 25°C. This value was close to the theoretical one.

An all-solid-state lithium battery, presented in Ref. [9], was assembled with sulfur as positive electrode materials and with an inorganic solid electrolyte a-$60Li_2S{\cdot}40SiS_2$ (mol %). The investigated all-solid-state lithium battery showed successive reversibility with large specific capacity. The specific capacity calculated on the basis of the mass of sulfur was more than $1000\,mAh\,g^{-1}$.

According to Ref. [10], amorphous TiS_3 active materials were prepared using a high-energy ball mill and applied to all-solid-state cells with Li_2S-P_2S_5 solid electrolyte. The solid-state cell operated as a rechargeable lithium battery with a high capacity of $400\,mAh\,g^{-1}$ at room temperature. The cell exhibited a higher capacity than a solid-electrolyte cell with crystalline TiS_2 and showed better cyclability than a liquid-electrolyte cell with crystalline TiS_3.

According to Ref. [11], composite positive electrodes for all-solid-state lithium batteries were prepared by mixing $LiCoO_2$ and Li_2S-P_2S_5 solid electrolytes particles. Two different sizes of solid electrolyte particles were used. It was found that the charge-discharge reaction proceeded uniformly in the positive electrode with smaller solid electrolyte particles.

During studies described in Ref. [12], a lithium superionic conductor of $Li_7P_{2.9}Mn_{0.1}S_{10.7}I_{0.3}$ was used as a solid electrolyte via high-energy milling. The $Li_7P_{2.9}Mn_{0.1}S_{10.7}I_{0.3}$ glass-ceramic possessed a high ionic conductivity of $5.6\,mS\,cm^{-1}$ at room temperature and a wide voltage stability up to 5 V vs. Li/Li^+. Sulfur-based composites were prepared by mixing sulfur powder, carbon black, and the solid electrolyte and utilized as the cathode for all-solid-state LSBs. The sulfur composite cathode showed a large capacity of

796 mAh g^{-1} at 0.05 C and much better cycling stability than the counterpart with organic liquid electrolyte.

According to Ref. [13] the interface resistance and stress/strain of sulfur cathodes were significantly reduced by conformal coating ≈2 nm sulfur (S) onto reduced graphene oxide (rGO). A Li-S full cell consisting of an rGO@S-$Li_{10}GeP_2S_{12}$-acetylene black (AB) composite cathode was evaluated. At 60°C, the all-solid-state Li-S cell showed a similar electrochemical performance as in liquid organic electrolyte, with high rate capacities of 1525.6, 1384.5, 1336.3, 903.2, 502.6, and 204.7 mAh g^{-1} at 0.05, 0.1, 0.5, 1.0, 2.0, and 5.0 C, respectively. It could maintain a high and reversible capacity of 830 mAh g^{-1} at 1.0 C for 750 cycles. The uniform distribution of the rGO@S nanocomposite in the $Li_{10}GeP_2S_{12}$-AB matrix generated uniform volume changes during lithiation/delithiation, significantly reducing the stress/strain, thus extending the cycle life. Minimization of the stress/strain of solid cells allowed a long cycle life of all-solid-state Li-S batteries.

REFERENCES

[1] P. López-Aranguren, N. Berti, A.H. Dao, J. Zhang, F. Cuevas, M. Latroche, C. Jordy, An all-solid-state metal hydride—sulfur lithium-ion battery, J. Power Sources 357 (2017) 56–60.

[2] S. Das, P. Ngene, P. Norby, T. Vegge, P.E. De Jongh, D. Blanchard, All-solid-state lithium-sulfur battery based on a nanoconfined $LiBH_4$ electrolyte, J. Electrochem. Soc. 163 (9) (2016) A2029–A2034, https://doi.org/10.1149/2.0771609jes.

[3] K.(K.). Fu, Y. Gong, G.T. Hitz, D.W. McOwen, Y. Li, S. Xu, Y. Wen, L. Zhang, C. Wang, G. Pastel, J. Dai, B. Liu, H. Xie, Y. Yao, E.D. Wachsman, L. Hu, Three-dimensional bilayer garnet solid electrolyte based high energy density lithium metal–sulfur batteries, Energy Environ. Sci. 10 (2017) 1568–1575.

[4] H. Nagata, Y. Chikusa, All solid state lithium–sulfur battery with high energy and power densities at the cell level, Energy Technol. 4 (4) (2016) 484–489.

[5] H. Nagata, Y. Chikusa, An all-solid-state lithium–sulfur battery using two solid electrolytes having different functions, J. Power Sources 329 (2016) 268–272.

[6] H. Nagata, Y. Chikusa, All-solid-state lithium–sulfur batteries using a conductive composite containing activated carbon and electroconductive polymers, Chem. Lett. 43 (8) (2014) 1335–1336.

[7] Y. Fujii, Y. Suto, A. Miura, M. Higuchi, K. Tadanaga, Development of all-solid-state lithium secondary batteries using $NiPS_3$ electrode and Li_2S-P_2S_5 solid electrolyte, Chem. Lett. 45 (6) (2016) 652–654.

[8] H. Nagata, Y. Chikusa, An all-solid-state sodium–sulfur battery operating at room temperature using a high-sulfur-content positive composite electrode, Chem. Lett. 43 (8) (2014) 1333–1334.

[9] N. Machida, T. Shigematsu, An all-solid-state lithium battery with sulfur as positive electrode materials, Chem. Lett. 33 (4) (2004) 376–377.

[10] A. Hayashi, T. Matsuyama, A. Sakuda, M. Tatsumisago, Amorphous titanium sulfide electrode for all-solid-state rechargeable lithium batteries with high capacity, Chem. Lett. 41 (9) (2012) 886–888.

[11] M. Otoyama, Y. Ito, A. Hayashi, M. Tatsumisago, Investigation of state-of-charge distributions for $LiCoO_2$ composite positive electrodes in all-solid-state lithium batteries by Raman imaging, Chem. Lett. 45 (7) (2016) 810–812.
[12] R.-C. Xu, X.-H. Xia, S.-H. Li, S.-Z. Zhang, X.-L. Wang, J.-P. Tu, All-solid-state lithium–sulfur batteries based on a newly designed $Li_7P_{2.9}Mn_{0.1}S_{10.7}I_{0.3}$ superionic conductor, J. Mater. Chem. A 5 (2017) 6310–6317.
[13] X. Yao, N. Huang, F. Han, Q. Zhang, H. Wan, J.P. Mwizerwa, C. Wang, X. Xu, High-performance all-solid-state lithium–sulfur batteries enabled by amorphous sulfur-coated reduced graphene oxide cathodes, Adv. Energy Mater., Nano Lett. 16 (11) (2016) 7148–7154.

CHAPTER FIFTEEN

Polysulfide Flow Batteries

A popular type of polysulfide flow cell is the bromine-polysulfide redox flow cell (Br/polyS RFC).

According to Ref. [1], the bromine-polysulfide redox flow cell is based on the use of the Br^-/Br_2 and S_2^{2-}/S_4^{2-} redox couples [2–4], which allow a high theoretical cell potential (1.5 V) to be achieved. During the charge process, bromide ions are oxidized at the surface of a graphite or carbon felt electrode [5] and form the tribromide ion (Fig. 56F), while polysulfide ions are reduced to sulfides at the porous nickel sulfide or nickel foam electrode [2].

Since all the electrochemically active species in the system are anions, a cation selective membrane (usually Nafion) is used to separate the catholyte and anolyte streams to prevent the bromine from directly reacting with sulfur anions [6]. Typically, 1 M NaBr solution saturated with bromine and 2 M Na_2S are used as the positive and negative electrolytes, respectively. Therefore, to maintain the electrical balance within the electrochemical cell, the charge is carried by sodium ions. The maximum achievable energy efficiency in this system was ~77% at the current density of 40 mA cm^{-2} [2].

The major challenge of this type of redox flow cell is the electrolyte crossover, which can lead to undesirable precipitation of sulfur species as well as formation of hazardous hydrogen sulfide and bromine [2,4].

As stated in the review [7], bromine-polysulfide redox flow battery (RFB) systems were introduced by Remick and Ang in 1984 [8] and developed by Regenesys Technologies (UK) at Innogy plc between 1991 and 2004 [9–13]. A 1 MW pilot plant at Aberthaw power state in South Wales was investigated with a round-trip efficiency of 60%–65% and energy density of 20–30 Wh. The numerical model of the Regenesys Technologies Ltd. pilot-scale bromine/polysulfide redox flow battery system, which allowed optimization of some operational parameters and predicted an energy efficiency of ~60%, was described in Refs. [14, 15].

The construction of 15 MW/120 MWh demonstration plants started at Little Barford in the early 2000s [10,11]. Regenesys Technologies was acquired by VRB Power Systems Inc. in 2004 and was subsequently acquired by Prudent Energy (China) in 2009.

Next-generation Batteries with Sulfur Cathodes
https://doi.org/10.1016/B978-0-12-816392-4.00015-3

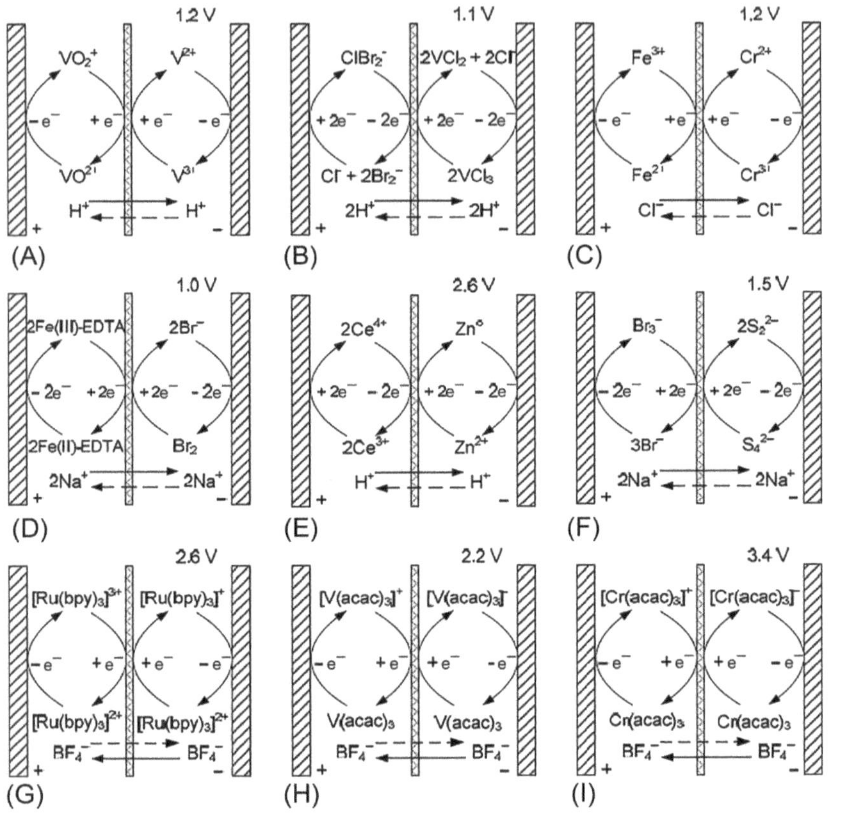

Fig. 56 Schematic of charge transfer in different redox flow systems. (A) All vanadium, (B) vanadium/bromine, (C) iron/chromium, (D) Fe-EDTA/bromine, (E) zinc/cerium, (F) bromine/polysulfide, (G) nonaqueous ruthenium, (H) nonaqueous vanadium, and (I) nonaqueous chromium [4].

The bromine-polysulfide RFB employed NaBr electrolyte in the positive half-cell and Na_2S_4 electrolyte in the negative half-cell. During the charging process, Br^- ions were oxidized to Br_2 and complexed as Br_3^- ions at the positive electrode while the sulfur, presented as the soluble polysulfide S_4^{2-} anion, was reduced to sulfide S_2^{2-} ions at the negative electrode. The positive and negative half-cells were separated by a cation exchange membrane to prevent the sulfur anions reacting directly with bromine and the electrical balance was achieved by the transport of Na+ ions across the membrane [2,10]. The electrode reactions were complex, especially at the negative electrode, but might be simplified to:

Negative electrode.

$$S_4^{2-} + 2e^- \underset{\text{discharge}}{\overset{\text{charge}}{\rightleftarrows}} 2S_2^{2-} \tag{23}$$

Positive electrode.

$$3Br^- - 2e^- \underset{\text{discharge}}{\overset{\text{charge}}{\rightleftarrows}} Br_3^- \tag{24}$$

During discharge, the sulfide ion was the reducing agent and the tribromide ion the oxidizing species. In contrast to other systems using bromide salts, no complexing agent was required.

The open-circuit cell voltage was reported to be 1.74 V in a fully charged state and 1.5 V at a 50% state of charge when operated at 25°C in 1 M NaBr and 2 M Na_2S_4, which were separated by a Nafion 125 membrane, using graphite as the positive electrode and porous nickel sulfide as the negative electrode [8].

The bromine-polysulfide RFB technology was scaled up at three sizes, in which 60, 120, and 200 bipolar electrodes, respectively, were assembled into a modular stack using a filter-press type of assembly. The nominal power ratings of the stacks were 5, 20, and 100 kW, respectively. A key feature of the modular technology was the use of carbon-polyolefin composite electrodes, enabling the electrodes to be welded to the polyolefin cell frames using laser-based production engineering techniques [11].

To improve the efficiency and performance of the bromine-polysulfide RFB, various electrode materials, including activated carbon, carbon/graphite felt, nickel foam, and metal-coated graphite felt, were investigated [5,16–18]. Using cobalt-coated graphite felt as the negative electrode and graphite felt as the positive one, the battery achieved a stable energy efficiency of up to 81% over 50 cycles (600 h) at a current density of 40 mA cm^{-2}.

There are several technical challenges facing the bromine-polysulfide technology:

(i) there is a risk of cross-contamination of the electrolytes due to crossover of the ionic species through the membrane;

(ii) long-term transport of species across the membrane leads to the need for analytical monitoring and chemical treatment of the electrolytes, increasing cost and complexity of the process plant;

(iii) this system also encountered the problem of build-up of sulfur species (e.g., S^{2-} and/or HS^{-}) either on the electrodes or membrane and therefore over extended cycling there is a net loss of sulfur from the system;

(iv) mixing of the electrolytes can generate heat and toxic gases such as Br_2 and H_2S.

Fig. 57 shows the plots of the cell voltage vs. time of the charge-discharge cycle of the four systems: all-vanadium (30 mA cm^{-2} for 2 h), zinc-bromine (15 mA cm^{-2} for 10 h), bromine-polysulfide (40 mA cm^{-2} for 30 min), and the undivided zinc-cerium system (20 mA cm^{-2} for 30 min).

An interesting prototype of a lithium/polysulfide semisolid rechargeable flow battery was presented in Ref. [19]. The feasibility of a semisolid flow battery with polysulfide as a catholyte was demonstrated, which gave a power density of 1.823 mW cm^{-2} at 4 mA cm^{-2}.

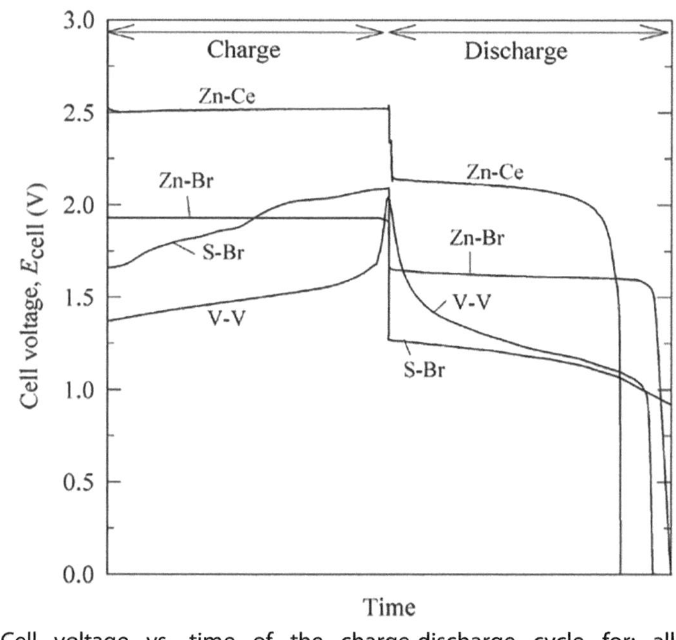

Fig. 57 Cell voltage vs. time of the charge-discharge cycle for: all-vanadium (30 mA cm^{-2} for 2 h), zinc-bromine (15 mA cm^{-2} for 10 h), bromine-polysulfide (40 mA cm^{-2} for 30 min), and the undivided zinc-cerium system (20 mA cm^{-2} for 30 min). *(Taken from P. Leung, X. Li, C. Ponce de León, L. Berlouis, C.T.J. Low, F.C. Walsh, Progress in redox flow batteries, remaining challenges and their applications in energy storage, RSC Adv. 2 (2012) 10125–10156.)*

Because the diffusion of polysulfides from the confined matrixes is unavoidable, and is essential for the sulfur cathode, polysulfide as a cathode active material for the Li/PS battery was proposed [20–22].

A Li/PS battery with an appropriate amount of polysulfide infiltrating into carbon paper was reported on, and a proof-of-concept trial was carried out by using a narrow voltage window to avoid the volume expansion for polysulfide's further large-scale energy storage [20,23].

In Ref. [19] a semisolid polysulfide flow cell (SSPFC) was introduced, employing polysulfides but in flow-through mode, giving a more realistic response. As shown in Fig. 58, a laboratory-scale cell was designed and constructed employing carbon as the reaction matrix of the cathode with polysulfides as a catholyte continually flowing through its surface.

Lithium plates are employed as anode and lithium polysulfide solution is used as a catholyte with carbon as the reaction matrix of the cathode. A Li/Li symmetric cell with blank electrolyte (BE) stored at 25°C was used to investigate the solid electrolyte interface (SEI) growth on the Li electrode. The passivation resistance of the Li electrode gradually increases with the storage time. In addition, the effect of the $LiNO_3$ additive was examined, which further demonstrated that an insoluble oxide passivation layer was generated on the surface of the Li electrodes [24,25]. The passivation layer not only

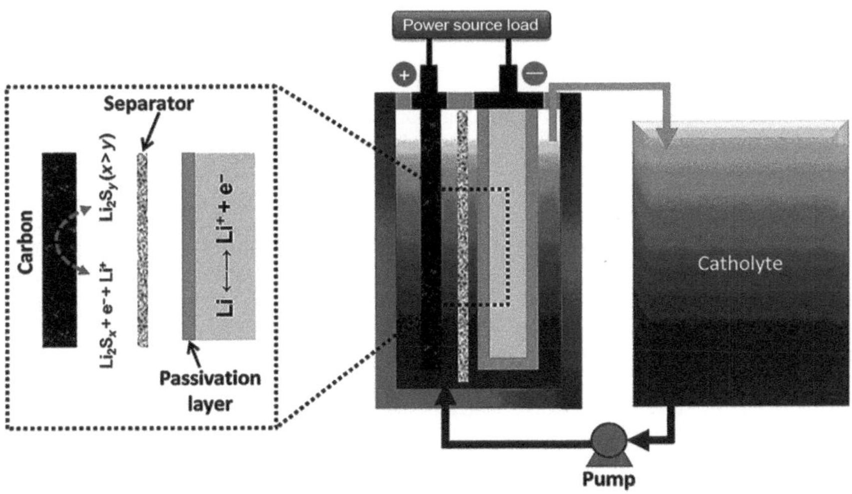

Fig. 58 Schematic diagram of the lithium/polysulfide semisolid flow cell. *(Taken from K. Dong, S. Wang, J. Yub, A lithium/polysulfide semi-solid rechargeable flow battery with high output performance, RSC Adv. 4 (2014) 47517–47520, doi: 10.1039/C4RA08413F (Communication).)*

protects the lithium from the chemical reaction with polysulfides and the "shuttle phenomenon" but also makes it possible to build the SSPFC. This protection effectively improves the utilization of the active materials on both sides of the electrodes and the cycling efficiency, avoiding the tremendous waste of energy during the charging progress in practical applications for the SSPFC.

REFERENCES

[1] V. Pupkevich, Scale-Up and Study of the BioGenerator, 2014, p. 2247 (Electronic Thesis and Dissertation Repository). https://ir.lib.uwo.ca/etd/2247.
[2] C. Ponce de León, A. Frias-Ferrer, J. Gonzalez-Garcia, D.A. Szanto, F.C. Walsh, Redox flow cells for energy conversion, J. Power Sources 160 (2006) 716–732.
[3] T. Nguyen, R.F. Savinell, Flow batteries, Electrochem. Soc. Interface (Fall) (2010) 54–56.
[4] A.Z. Weber, M.M. Mench, J.P. Meyers, P.N. Ross, J.T. Gostick, Q. Liu, Redox flow batteries: a review, J. Appl. Electrochem. 41 (2011) 1137–1164.
[5] H. Zhou, H. Zhang, P. Zhao, B. Yi, A comparative study of carbon felt and activated carbon based electrodes for sodium polysulfide/bromine redox flow battery, Electrochim. Acta 51 (2006) 6304–6312.
[6] Z. Yang, J. Zhang, M.C.W. Kintner-Meyer, X. Lu, D. Choi, J.P. Lemmon, J. Liu, Electrochemical energy storage for green grid, Chem. Rev. 111 (2011) 3577–3613.
[7] P. Leung, X. Li, C. Ponce de León, L. Berlouis, C.T.J. Low, F.C. Walsh, Progress in redox flow batteries, remaining challenges and their applications in energy storage, RSC Adv. 2 (2012) 10125–10156.
[8] R.J. Remick, P.G.P. Ang, US 1984/4485154, 1984.
[9] R. Zito, US Pat. 5612148, 1997.
[10] A. Price, S. Bartley, S. Male, G. Cooley, A novel approach to utility scale energy storage [regenerative fuel cells], Power Eng. J. 13 (1999) 122–129.
[11] F.C. Walsh, Electrochemical technology for environmental treatment and clean energy conversion, Pure Appl. Chem. 73 (2001) 1819–1837.
[12] P.J. Morrissey, P.J. Mitchell, S.E. Male, Pat. Pub. No.: WO/2001/003221, 2001.
[13] P.J. Morrissey, N.J. Ward, GB Pat. 2374722, 2002.
[14] D.P. Scamman, G.W. Reade, E.P.L. Roberts, Numerical modelling of a bromide–polysulphide redox flow battery. Part 1: modelling approach and validation for a pilot-scale system, J. Power Sources 189 (2) (2009) 1220–1230.
[15] D.P. Scamman, G.W. Reade, E.P.L. Roberts, Numerical modelling of a bromide-polysulphide redox flow battery. Part 2: evaluation of a utility-scale system, J. Power Sources 189 (2) (2009) 1231–1239.
[16] S.H. Ge, B.L. Yi, H.M. Zhang, Study of a high power density sodium polysulfide/bromine energy storage cell, J. Appl. Electrochem. 34 (2004) 181–185.
[17] P. Zhao, H.M. Zhang, H.T. Zhou, B.L. Yi, Nickel foam and carbon felt applications for sodium polysulfide/bromine redox flow battery electrodes, Electrochim. Acta 51 (2005) 1091–1098.
[18] H. Zhou, H. Zhang, P. Zhao, B. Yi, Novel cobalt coated carbon felt as high performance negative electrode in sodium polysulfide/bromine redox flow battery, Electrochem. Commun. 74 (2006) 296–298.

[19] K. Dong, S. Wang, J. Yub, A lithium/polysulfide semi-solid rechargeable flow battery with high output performance, RSC Adv. 4 (2014) 47517–47520, https://doi.org/10.1039/C4RA08413F (Communication).
[20] Y. Yang, G. Zheng, Y. Cui, A membrane-free lithium/polysulfide semi-liquid battery for large-scale energy storage, Energy Environ. Sci. 4 (2013) 1552–1558.
[21] Y. Fu, Y.-S. Su, A. Manthiram, Highly reversible lithium/dissolved polysulfide batteries with carbon nanotube electrodes, Angew. Chem. 125 (2013) 7068–7073.
[22] S.S. Zhang, J.A. Read, A new direction for the performance improvement of rechargeable lithium/sulfur batteries, J. Power Sources 200 (2012) 77–82.
[23] M. Agostini, D.-J. Lee, B. Scrosati, Y.K. Sun, J. Hassoun, Characteristics of Li_2S_8-tetraglyme catholyte in a semi-liquid lithium–sulfur battery, J. Power Sources 265 (2014) 14–19.
[24] D. Aurbach, E. Pollak, R. Elazari, G. Salitra, C.S. Kelley, J. Affinito, On the surface chemical aspects of very high energy density, rechargeable Li–sulfur batteries, J. Electrochem. Soc. 156 (2009) A694–A702.
[25] Y.V. Mikhaylik, US Pat. 2008/0193835 A1, 2008.

CHAPTER SIXTEEN

Modeling of Batteries

According to Ref. [1], there are a wide variety of battery models with varying degrees of complexity. They describe various aspects of battery behavior, from design and performance estimation to circuit simulation.

Electrochemical models [2–5] are used to optimize the physical design of batteries. They characterize the fundamental mechanisms of power generation and relate battery design parameters with macroscopic (e.g., battery voltage and current) and microscopic (e.g., concentration distribution) information. They are complex, time-consuming, and involve a system of coupled, time-variant, spatial, partial differential equations [4]. Their solution requires complex numerical algorithms and battery-specific information, usually difficult to obtain.

Mathematical models described in Refs. [6–11] are rather too abstract but are still useful to system designers. They adopt empirical equations or stochastic approaches [7] to predict system-level behavior, such as battery runtime, efficiency, or capacity. Mathematical models do not give any I-V information useful for circuit simulation and optimization. Most mathematical models are appropriate only for specific applications and provide inaccurate results on the order of 5%–20% error. For example, the maximum error of Peukert's law predicting runtime can be more than 100% for time-variant loads [9].

Electrical models [12–24], with accuracy between electrochemical and mathematical models (around 1%–5% error), are electrical equivalent models using a combination of voltage sources, resistors, and capacitors, for codesign and cosimulation with other electrical circuits and systems. Electrical models are intuitive, useful, and easy to handle, especially when they can be used in circuit simulators and alongside application circuits. There are many electrical battery models, from lead-acid to polymer lithium-ion (Li-ion) batteries. Most of these electrical models fall under three basic categories: Thevenin- [13–19], impedance- [20, 21], and runtime-based models [12, 22, 23].

16.1 MODELS OF LEAD-ACID AND LITHIUM-ION BATTERIES

It is rather commonly known that in the fully charged state of a lead-acid battery, the negative electrode consists of lead and the positive electrode

Next-Generation Batteries With Sulfur Cathodes
https://doi.org/10.1016/B978-0-12-816392-4.00016-5

lead dioxide, with the electrolyte of concentrated sulfuric acid. Overcharging with high charging voltages generates oxygen and hydrogen gas by electrolysis of water, which is lost to the cell. In the discharged state both the positive and negative electrodes become lead(II) sulfate ($PbSO_4$), and the electrolyte loses much of its dissolved sulfuric acid and becomes primarily water. The discharge process is driven by the conduction of electrons from the negative plate back into the cell at the positive plate in the external circuit. Lead sulfate, however, after a certain time changes into a crystalline state (electric insulator), thus causing a decrease in battery capacity. In such a situation, charging the battery is impossible, as the crystallized lead sulfate does not take part in chemical processes.

The models of a lead-acid battery are shown, for example, in Refs. [25, 26]. The model of the lithium-ion battery was discussed, among others, in Refs. [1, 7, 26–34].

Some measurement techniques applicable to lithium-ion cells can be carefully used also for lithium-sulfur cells and others. However, one should always consider the effect of physical and chemical phenomena on changes in the basic parameters of cells.

According to the review presented in Ref. [35], capacity, internal resistance, and self-discharge are the main basic parameters determining the performance of lithium ion batteries. For a given battery voltage and weight, the specific energy of a battery is determined by its capacity, while the internal resistance limits its specific power.

Under high current loads the heat evolution of the battery and energy efficiency are primarily determined by its internal resistance.

Resistance is generally measured by applying a voltage to the device under test and measuring the resulting current or by applying a current and measuring the resulting voltage. Interpretation of the measurement result is easy, when a device under test is governed by Ohm's law.

The definition of "internal resistance" for non-LTI (linear time invariant) systems is complex and unequivocal. While methods based on charge/discharge define "large signal resistance," methods like impedance spectroscopy define "small signal resistance." These two definitions are different for batteries (like other non-LTI systems) and could relate to each other using systems theory for nonlinear systems.

According to Ref. [36] the knowledge of the battery's internal resistance is needed for dimensioning the battery system, for selecting and comparing cells, for energy efficiency calculations, for dimensioning the cooling system of the battery, and for power estimation. Also, a proper design of the cooling system is needed for safety and lifetime of a battery system [37].

A battery is a case of a complex load, showing capacitive, ohmic, and inductive behavior. Its behavior also depends on the measurement frequency and amplitude. Therefore, it is difficult to measure the internal resistance of a single cell or battery, which are of a nonlinear and time-dependent system.

Determining the internal resistance of a battery is difficult because the results of the measurement are not only governed by the ohmic behavior of the device but also by its capacitive and inductive behavior. There is also additional nonlinearity, such as temperature dependence and time-variant behavior of the battery under test due to the specific measurement procedure.

According to Ref. [38], the internal resistance of the cell is also a function of the method of determination. The experimental conditions also influence the outcome of its measurements. Thus, sophisticated measurement procedures should be used for measuring the resistive part of a complex system. Such procedures use the frequency dependency of the battery under test.

According to Refs. [38, 39] the cell voltage U_{cell} under load depends on the open circuit voltage, the overvoltages caused by concentration polarization η_{diff} and charge transfer polarization $\eta_{ch,tr}$, and the voltage drop caused by internal ohmic resistance R_i:

$$U_{cell} - U_{ocv} - \eta_{diff} - \eta_{ch,\,tr} - IR_i \tag{25}$$

The internal resistance of a battery is usually its characteristic parameter but is not a simple ohmic resistance and depends on the method used for its determination, the state of charge of the battery and on its temperature. The "concentration polarization" and "charge transfer polarization" effects should be separated for voltage depression caused by ohmic internal resistance.

According to Ref. [40], the ohmic resistance also encompasses the resistivity of the components of the battery such as the active material of the anode and the cathode, the current collectors, and the electrolyte.

The internal resistance depends on material contacts, electrode geometry, and internal design aspects. The effect of this part of the total effective resistance can be observed during the first few milliseconds after start of battery discharging or recharging [41].

According to Refs. [38, 42, 43] the total effective resistance can be subsequently calculated by dividing the change in voltage by the change in current.

The magnitude of R_i is mainly determined by the processes at the interface between active material and electrolyte, the electrolyte conductivity, and the loss by the purely ohmic content of the supporting and conductive elements (electrodes, active mass, connectors, and tabs).

The internal resistance can be calculated from the current-voltage relationship given by two points in a discharge curve, similarly to the case of lead acid batteries [44, 45]:

$$R_i = \frac{U_1 - U_2}{I_2 - I_1} \tag{26}$$

The change of the voltage is observed during a cell operation in three possible states as defined by Kobayashi et al. [46]:
- during discharge of cell,
- during recharge of cell,
- during the interruption of current flow (charge/discharge).

According to Ref. [38], the direct current method can be used only for pure ohmic elements during the discharge or recharge of the battery.

The other method for calculating the cell internal resistance is based on the determining the Joule heat loss caused by internal resistance during the cell operation. A calorimeter can be used for measurement of heat from chemical reactions or physical changes [36, 46, 47] even for high-power cells with low heat generation.

The chemical reactions in the cells may be exothermic or endothermic [46, 47].

According to Ref. [47], in the case of lithium-ion batteries the reversible heat effect during charging, caused by intercalation of lithium ions to the lattice, is initially endothermic, then turns to slightly exothermic during most of the charging cycle. During discharge, the reaction is the reverse. Under high current load conditions, the Joule heating effect for lithium-ion cells is greater than the reversible heat effect; therefore during the determination of internal resistance by calorimeter measurements, those effects were omitted.

If a battery is cycled symmetrically around a given state of charge (SoC), the reversible heat effect is cancelled out, and only Joule's law governs heat dissipated by the battery. In such a case, internal reactance of the battery can be determined by Joule's heating with a calorimeter.

Heat dissipated by the cell can be calculated by the temperature increase, with determination of the heat capacity for the whole set-up made by injection of a known amount of heat into the system by the calibration heater.

For the calculation of the value of R_i from the known values of cell capacity C_{cell}, heat capacity of the calorimeter C_{cal}, temperature change ΔT, and amount of heat delivered into the system Q one needs to determine the temperature change during the charge/discharge cycle of the cell by calorimeter measurements with a substance ("calorimeter liquid") and while using a cell with heat capacities obtained from previous tests.

Internal resistance of the cell can be calculated from Eq. (27):

$$R_i = \frac{(C_{cal} + C_{cell}) \times (T_2 - T_1)}{\int_0^t (I(t))^2 dt} \tag{27}$$

In the complex case of the cell, the internal resistance can be determined via the alternating current (AC) resistance. The AC resistance of a battery can be measured by supplying the battery with a small alternating current $I(t)$ ripple that generates a small AC voltage $U(t)$ response, e.g., with a constant frequency of 1 kHz. Electrical impedance (complex resistance) is the relative ratio of the variation in current with the variation in voltage, describing the relative amplitudes of the voltage and current (like at the DC internal resistance), and the relative phases. In comparison to the direct current (DC) method (switch in/switch off), the AC method is additionally defined by the phase angle [38, 48]:

$$Z = \frac{U(t)}{I(t)} = \frac{U_{max} \exp(\omega t + \varphi_U)}{I_{max} \exp(\omega t + \varphi_I)} \tag{28}$$

The measured electrical impedance depends on the frequency of AC current used for the measurements. Different devices measuring the electrical impedance can create a significant variance of the results. The AC current method is suitable for complex resistances.

The complex cell internal resistance can be determined using electrochemical impedance spectroscopy (EIS). During this, a small amplitude AC signal is also applied to the cell.

According to Refs. [41, 48] EIS can provide the following information on the examined cell: corrosion rate, electrochemical mechanisms and reaction kinetics, detection of localized corrosion, battery life, and internal resistance (impedance).

During modeling the EIS can be useful in the equivalent circuit [49] treating an ohmic resistance (R_i) as a function of the battery contacts impedance, the inter- and intracell connections, the electrodes and the electrolyte, the SoC of the battery, the battery aging history, and the battery temperature. R_{t1} and C_{d1} are the charge-transfer resistance and the interfacial capacitance (including the double layer capacitance and associated capacitive components due to adsorption, passive films, etc.) for the first electrode, and R_{t2} and C_{d2} are the charge transfer resistance and interfacial capacitance for the second electrode.

The EIS testing described in Ref. [50] showed the budget cells to have internal impedances several times higher than other cells, leading to notably increased heat generation and a significantly reduced cell efficiency. Differential capacity analysis found this high internal resistance to notably impede lithium intercalation processes.

Internal resistance in the case of the lithium-ion cell was measured by the following methods [35]:

- VDA current step method,
- ISO current step method,
- Similar to DIN EN 62391-1,
- Current-off method,
- Switching current method,
- AC internal resistance,
- Impedance spectroscopy,
- Energy loss method,
- Quasi-adiabatic calorimeter.

It was found in Ref. [35] that a feigned resistance occurs by charging or discharging the battery when the internal resistance is determined by the voltage drop of long and high current charge or discharge pulses. Reduction of pulse current and pulse duration was determined to reduce the influence of discharge and charge on creating a feigned resistance. Reducing the pulse amplitude leads to unacceptable large measurement uncertainties. Such uncertainties can be kept in an acceptable range by using high current pulses and using the voltage just after the start of the current pulse used for calculating the internal resistance. A delay of 100 ms between start of the current pulse and voltage measurement enables a stable current output from the power stage of typical battery testing systems. The time period of 100 ms is short enough to prevent unintended change of charge of the battery, so the feigned resistance is negligible. Extrapolation to zero delay may be used to improve measurement accuracy, but within the accuracy of the equipment used in this study an improvement cannot be stated clearly.

The methods using steps give the same results, within measurement accuracy, if the outcome of the measurement is not falsified by charging or discharging the cell. The outcome of the measurement is independent of the sign of current. It is also independent if the voltage difference at the beginning or the end of the current pulse is used to calculate internal resistance of the cell. A direct switch, from charging to discharging, gives the same result as the other step methods. The internal resistance of the cell was also determined by measurement of the waste energy caused by high current cycling. Waste energy was measured by the difference of charge and discharge energy as well as by determining the dissipated heat of the cell using a calorimeter. Both methods give the same results for internal resistance, within measurement range. It was found that both energy loss methods give the same results for internal resistance as do current step methods (without change of charge). This is a clear hint that reversible heat effects of the cell are cancelled out if a symmetrical current profile is used. Because two measurement principles give the same results for internal resistance, it is expected that the values determined with these methods provide reliable information about the pure ohmic resistance of a cell. The method of measuring AC resistance at fixed frequency gives fast results. Because each cell type shows individual frequency dependence of impedance, AC resistance measured on different types of cells cannot be generally used for cell comparison and benchmarking. Therefore, this method is only suitable for measuring and comparing internal resistance of the same type of cell, e.g., for quality screening.

At frequencies about 1 kHz, AC resistance for different cells with comparable size has similar frequency dependence. This also could be used for comparing different cells. Cells with higher small-signal resistance have higher large-signal resistance. Comparison between values found with AC methods with the values found by step methods is not easy, because of the complex electrochemical nature of the cell and the broad range of frequencies found in the frequency domain. Since the fast steps and the energy loss methods provide the same results, these methods are recommended for the measurement of internal resistance for designing the battery system, selecting and comparing cells, energy efficiency calculation, dimensioning the cooling system of the battery, and performing power estimation. Determining the internal resistance by the voltage drop after a short delay after the start of a current pulse is a fast and convenient method for the measurement of internal resistance during storage tests, cell benchmarks and other highly automated cell screening purposes. Since values generated by this method are conform to the values from energy loss methods, they can replace the time-consuming energy loss methods.

16.2 MODELS OF LI-S BATTERIES

A critical analysis of Li-S cell models is presented in Ref. [51]. In Ref. [52], the first Li-S model for a polysulfide transfer study is described. This digitized model calculates the reduction potentials in high and low voltage plateau based on the Nernst equation, but omits the superconducting potential, electrolyte resistivity, and dissolution/precipitation reactions, which affects the operating voltage. A more detailed one-dimensional mechanical model (1D) was later described in Ref. [53] based on the Nernst-Planck equations for dilute solutions. The Kumaresan model includes multicomponent transport in porous cathodes and separators, charge transfer kinetics, dissolution/precipitation reactions, and changes in porosity and electrochemically active surface because of dissolution/precipitation. Although the model qualitatively reproduces some of the essential features of a typical Li-S discharge profile—such as two characteristic voltage plateaus and voltage drops between them—it also requires the introduction of many physical and phenomenological parameters that are not easy to obtain experimentally. Due to the complexity of the model, a sensitivity analysis of this model was carried out [54–56]. It was found that for the model to be able to predict the loading process, it is necessary to further develop a model for the precipitation/dissolution reaction. In Refs. [57, 58] a 1D Li-S model is presented based on a similar structure, which additionally includes the electrochemical bilayer, polysulfide transfer, and irreversible precipitation in the anode as a mechanism reducing capacity. In addition to charging/discharging curves, the electrochemical impedance spectrum was predicted at various discharge depths (DOD). Although the mechanistic models [53, 57, 58] can reproduce the overall discharge profile of the Li-S cell, they cannot capture more complex cell behavior, such as change in electrolyte resistance (R_s) during discharging. Variability of R_s together with DOD is a characteristic feature of Li-S cells observed in various chemical behaviors of electrolytes [54, 59–61]. As shown in Fig. 59A, during discharge R_s rises on a high plateau and reaches maximum during the transition between two plateau voltages; then it decreases on a low plateau. In addition, R represents most of the voltage drop on a high plateau, because low-frequency resistances—often related to the load—are significant only on the low plateau [61–63]. In mechanistic models, the predicted voltage drop caused by electrolyte resistance is from two to three orders smaller than the predicted voltage drop because of overpotential activation. Such model prediction, contrary to experimental data, is visible in simulated model curves obtained in Ref. [57]. This also disagrees with the predictions resulting from the Kumaresan model, of which nine are shown in Fig. 59B, where simulated R_s exhibits very different evolution along with DOD, compared to measurements. The R-profile at discharge is typically explained by

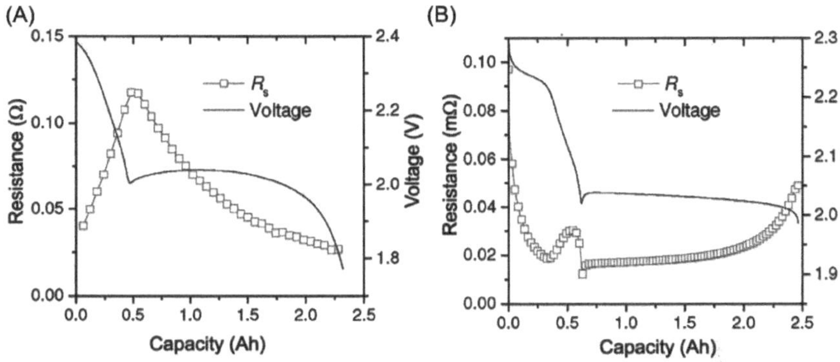

Fig. 59 (A) Electrolyte resistance (R_s) measured for the Li-S cell produced by Oxis Energy Ltd. EIS galvanostatic measurements were performed during the discharge of the cell with 0.34 A current, and the series resistance was separated when the *x* axis at high frequencies crossed Nyquist diagrams, for different discharge depths. (B) Simulated R_s from the Kumaresan model [53] for Li-S cells with similar energy efficiency; model parameters were taken from [56].

changing the concentration of electrolyte (or viscosity) caused by the dissolution and subsequent precipitation of the polysulfides. In fact, it was experimentally observed that the conductivity of the Li-S cells electrolyte strongly depends on the concentration of both the lithium salt [64] and lithium polysulfides [65]. At high concentrations of salt or polysulfide (typically $>1\,\mathrm{mol\,L^{-1}}$), increased ionic interactions reduce the conductivity of the electrolyte. Existing Li-S models, however, rely on the solution dilution theory in which ionic conductivity is independent of ion concentration.

In Ref. [51], it was proposed to introduce dependence of electrolyte conduction on concentration to recover experimentally documented voltage drop trends in Li-S cells during discharge. This feature is included in the mechanistic model that describes the electrochemical and precipitation reactions, the kinetics of charge transfer of electrodes, as well as morphological changes caused by the precipitation in the Li-S cell. Compared to the more detailed Kumaresan model, the discretized model does not consider the effects of mass transport and cargo location, and therefore cannot predict transport limitations. However, the sensitivity analysis [56] shows that mass transport does not significantly affect the predictions of the Kumaresan model, if the ion diffusion coefficients do not fall by more than an order of magnitude. Consequently, the discharge curves generated by the discretized model resemble those obtained in the Kumaresan model.

The advantage of the discretized approach is that the model requires fewer matching parameters and lower computing power. By introducing concentration-dependent electrolyte conductivity, correct voltage-drop trends due to electrolyte resistance and overpotential activation are obtained. In addition, there is often a neglected mechanism of potential decline in the low voltage plateau that results from the limited Li_2S precipitation rate.

Some electrochemical models have been developed for the Li-S cell [54–56, 66, 67]. A zero-dimensional model, described in Ref. [67], can predict the behavior of a Li-S cell during charge and discharge at a good level of accuracy. Such models are useful for understanding the complex behavior of a Li-S cell.

Due to its practical usefulness, the equivalent circuit network (ECN) modeling approach is very common among researchers and engineers. The original application of ECN models to lithium sulfur was carried out in works such as [68–71].

In Ref. [72] a simple but comprehensive mathematical model of the Li-S battery cell self-discharge based on the shuttle current was developed. The shuttle current values for the model parameterization were obtained from the direct shuttle current measurements. Furthermore, the battery cell DOD values were recomputed to account for the influence of the self-discharge and provide a higher accuracy of the model. Finally, the derived model was successfully validated against laboratory experiments at various conditions.

Accurate estimation of the battery states (SoC and state of health, or SoH) is possible using the Kalman filter family. They have been successfully applied to lithium-ion batteries [73–75].

In Ref. [76], extended Kalman filter (EKF), unscented Kalman filter (UKF), and particle filter techniques were used for real Li-S cell state estimation subject to automotive drive cycles. Closely related to these was a set of techniques from artificial intelligence based on adaptive neurofuzzy inference systems (ANFISs). In Ref. [77], an ANFIS structure was used to estimate a real Li-S cell's SoC based on real-time cell model parameterization. In this approach, parameters of an ECN cell model were used as an indicator of SoC in real time.

According to Ref. [78], one big challenge in state estimation and control of a Li-S cell is its flat open-circuit-voltage (OCV) curve vs. SoC. The Li-S cell SoC range can be divided into two parts, usually referred to as the high plateau (HP) and low plateau (LP). There is a breakpoint at around 75% SoC that determines the boundary between the two plateaus. Such transition caused by a sudden change in electrochemical reactions inside the Li-S cell, might shift slightly to the right or left under different discharge conditions [77].

The flat shape of the OCV curve at LP makes the system unobservable based on control theory [77]. This unique feature of the Li-S cell makes its control a very challenging task.

Another challenge in estimation and control of the Li-S cell is its high sensitivity to the temperature variation [69, 71]. Temperature and the Li-S cell's ohmic resistance are inversely proportional. Temperature also influences plateau change (the breakpoint between HP and LP regions). The results in Ref. [69] demonstrate that the HP region shrinks as the temperature increases. The breakpoint moves from 65% at 10°C to 85% at 50°C. The Li-S cell's ohmic resistance has a highly nonlinear relationship with SoC. Ohmic resistance is low at high SoC, increasing linearly at HP by charge depleting. There is a break-point at the end of HP where cell's ohmic resistance starts decreasing after that. At LP, Li-S cell's ohmic resistance almost has a parabolic shape with a minimum point in the middle [69]. Li-S cell's ohmic resistance has a highly nonlinear relationship with SoC. Ohmic resistance is low at high SoC, increasing linearly at HP by charge depleting. There is a breakpoint at the end of HP where the cell's ohmic resistance starts decreasing after that. At LP, the Li-S cell's ohmic resistance almost has a parabolic shape with a minimum point in the middle [69].

16.3 NA-S BATTERY MODELS

Several battery models for modeling the battery characteristics, e.g., Na-S, are shown in Ref. [79].

16.3.1 A Simple Battery Model

The most commonly used battery model is shown in Fig. 60, consisting of a constant internal resistance (R_0) and open circuit voltage (E_0), where V_0 is the final battery voltage [80]. Because the internal resistance of the Na-S battery is sensitive and changes with the temperature and depth of discharge, this model

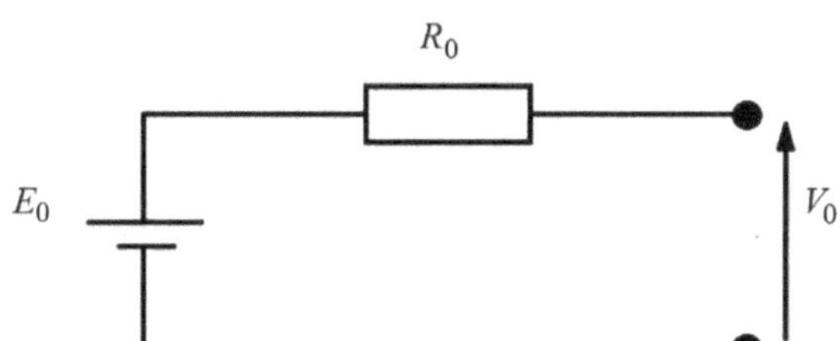

Fig. 60 A simple battery model [80].

is not suitable for modeling the Na-S battery, because it does not consider the different internal characteristics of the battery resistance in terms of discharge depth and temperature change. A simple model can be used only in simplified case studies, where it can be assumed that E_0 energy is unlimited.

16.3.2 Thevenin Battery Model

The second most-used battery model is the Thevenin model, which includes the ideal unloaded battery voltage (E_0), internal resistance (R), capacitance (C_0) and superheat resistance (R_0), as shown in Fig. 61 [80, 81].

When modeling the Na-S battery, the disadvantage of this model is that the elements are fixed, while in fact all element values are related to the battery conditions [65]. In addition, the internal voltage drop in the open circuit (decrease of EMF) in the Na-S battery is not considered.

16.3.3 Modified Battery Model

This modified battery model, shown in Fig. 62, is relatively simple, but at the same time meets most of the requirements for Na-S battery modeling.

This is a modified battery model based on the basic configuration of a simple battery model. It considers the nonlinear characteristics of the battery components during charging and discharging, as well as the internal resistance, which depends on the temperature changes and the depth of the battery

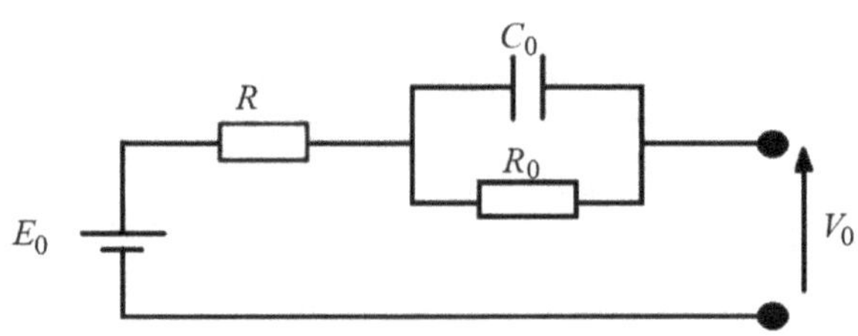

Fig. 61 Thevenin model of the battery [80, 81].

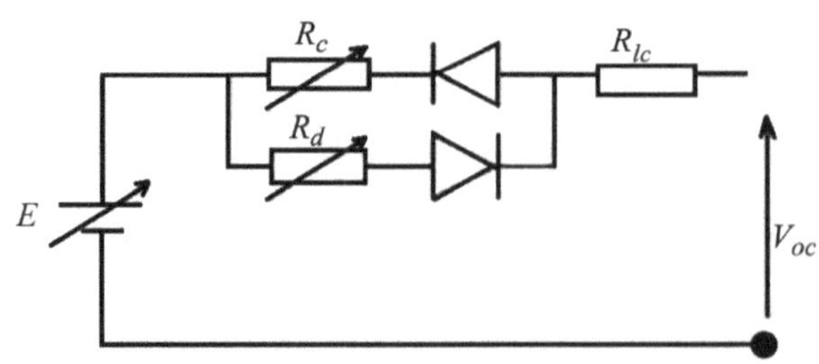

Fig. 62 Modified battery model [79].

discharge. Therefore, it is the most suitable for Na-S battery modeling using the EMTDC/PSCAD simulation tool. Based on the Na-S battery characteristics, the components of the modified battery model are described as follows.

16.3.3.1 Charging and Discharging Resistance, R_c *and* R_d

The internal resistance R_c and R_d is a function of temperature and depth of discharge. The purpose of introducing the diodes is that the Na-S battery has a different value of internal resistance during the loading and unloading operation.

16.3.3.2 Resistance During Cycles Loading-Discharging, R_{lc}

R_{lc} represents the increase in resistance due to deterioration, which is based on the number of cyclic resistances of charging and discharging.

16.3.3.3 Battery Circuit Voltage, E

Voltage E represents the EMF cell as a function of discharge depth. Based on the characteristics shown in Fig. 5, voltage E is expressed as follows:

$$E = E_0;\text{for DOD} \le 60\% \tag{29}$$

$$E = E_0 - k \cdot f;\text{for DOD} > 60\% \tag{30}$$

where:

k—constant obtained experimentally,

f—depth of discharge (%),

E_0—EMF when fully charged.

The modified battery model from Fig. 41 shows the performance of one Na-S cell.

The battery energy storage system uses battery modules that contain several cells connected in series and/or in parallel. The voltage behavior of such a battery module can be estimated and simulated by multiplying the cell's internal cell resistance and EMF by the number of cells, based on the cell's internal resistance data and performance information from the battery manufacturer.

16.3.3.4 Simulation Model in the PSCAD/EMTDC Program

Modeling and simulation of the Na-S battery is carried out using the PSCAD/EMTDC program, a standard simulation tool for studying the transient behavior of electrical networks. Using one of the built-in elements available in the PSCAD program, relevant data and parameters of the manufacturer are compiled in a reference table for viewing data. Consequently, the properties of the elements R_c, R_d, E, and R_{lc} are plotted and derived from the reference

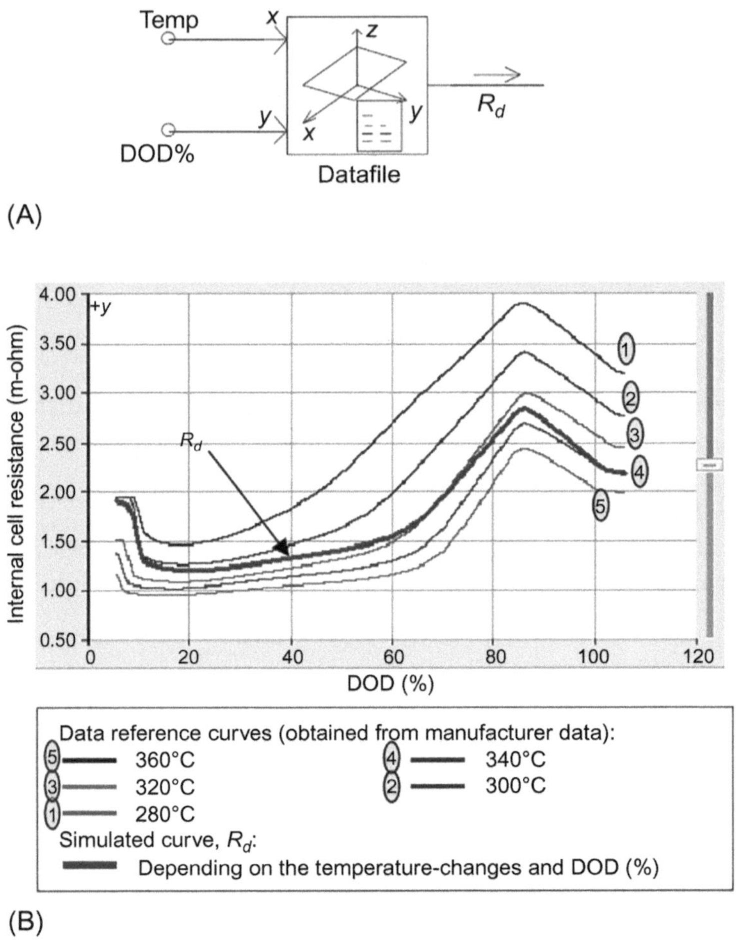

Fig. 63 (A) The *XYZ* component provided in the PSCAD program; (B) the simulated curve R_d as a function of temperature and DOD% [79].

data review table using linear interpolation or extrapolation, depending on the input value of discharge depth and temperature.

Fig. 63A shows the PSCAD component used, more precisely the *XYZ* component, which outputs the z value (curve R_d) based on input values x (temperature change) and y (discharge depth, %) using interpolation relative to reference files based on experimental data.

The details of this dependence of *XYZ* are illustrated in Fig. 63B, which shows the values of internal cell resistance for different release depths and temperatures.

REFERENCES

[1] M. Chen, G.A. Rincón-Mora, An accurate electrical battery model capable of predicting runtime and I–V performance, IEEE Trans. Energy Convers. 21 (2) (2006).
[2] L. Song, J.W. Evans, Electrochemical-thermal model of lithium polymer batteries, J. Electrochem. Soc. 147 (2000) 2086–2095.
[3] P.M. Gomadam, J.W. Weidner, R.A. Dougal, R.E. White, Mathematical modeling of lithium-ion and nickel battery systems, J. Power Sources 110 (2) (2002) 267–284.
[4] D.W. Dennis, V.S. Battaglia, A. Belanger, Electrochemical modeling of lithium polymer batteries, J. Power Sources 110 (2) (2002) 310–320.
[5] J. Newman, K.E. Thomas, H. Hafezi, D.R. Wheeler, Modeling of lithium-ion batteries, J. Power Sources 119–121 (2003) 838–843.
[6] M. Pedram, Q. Wu, Design considerations for battery-powered electronics, in: Proc. 36th Des. Autom. Conf., 1999, pp. 861–866.
[7] C.F. Chiasserini, R.R. Rao, Energy efficient battery management, IEEE J. Sel. Areas Commun. 19 (7) (2001) 1235–1245.
[8] R. Rynkiewicz, Discharge and charge modeling of lead acid batteries, in: Proc. Appl. Power Electron. Conf. Expo., vol. 2, 1999, pp. 707–710.
[9] D. Rakhmatov, S. Vrudhula, D.A. Wallach, A model for battery lifetime analysis for organizing applications on a pocket computer, IEEE Trans. VLSI Syst. 11 (6) (2003) 1019–1030.
[10] P. Rong, M. Pedram, An analytical model for predicting the remaining battery capacity of lithium-ion batteries, in Proc. Design, Automation, and Test in Europe Conf. and Exhibition (2003) 1148–1149.
[11] P.E. Pascoe, A.H. Anbuky, VRLA battery discharge reserve time estimation, IEEE Trans. Power Electron. 19 (6) (2004) 1515–1522.
[12] L. Benini, G. Castelli, A. Macci, E. Macci, M. Poncino, R. Scarsi, Discrete-time battery models for system-level low-power design, IEEE Trans. VLSI Syst. 9 (5) (2001) 630–640.
[13] Z.M. Salameh, M.A. Casacca, W.A. Lynch, A mathematical model for lead-acid batteries, IEEE Trans. Energy Convers. 7 (1) (1992) 93–98.
[14] M. Valvo, F.E. Wicks, D. Robertson, S. Rudin, Development and application of an improved equivalent circuit model of a lead acid battery, in: Proc. Energy Convers. Eng. Conf., vol. 2, 1996, pp. 1159–1163.
[15] M. Ceraolo, New dynamical models of lead-acid batteries, IEEE Trans. Power Syst. 15 (4) (2000) 1184–1190.
[16] S. Barsali, M. Ceraolo, Dynamical models of lead-acid batteries: Implementation issues, IEEE Trans. Energy Convers. 17 (1) (2002) 16–23.
[17] B. Schweighofer, K.M. Raab, G. Brasseur, Modeling of high power automotive batteries by the use of an automated test system, IEEE Trans. Instrum. Meas. 52 (4) (2003) 1087–1091.
[18] L. Gao, S. Liu, R.A. Dougal, Dynamic lithium-ion battery model for system simulation, IEEE Trans. Compon. Packag. Technol. 25 (3) (2002) 495–505.
[19] M.C. Glass, Battery electrochemical nonlinear/dynamic SPICE model, in: Proc. Energy Convers. Eng. Conf., vol. 1, 1996, pp. 292–297.
[20] S. Buller, M. Thele, R.W.D. Doncker, E. Karden, Impedance-based simulation models of supercapacitors and Li-ion batteries for power electronic applications, in: Conf. Rec. 2003 Ind. Appl. Conf., vol. 3, 2003, p. 159601600.
[21] P. Baudry, M. Neri, M. Gueguen, G. Lonchampt, Electro-thermal modeling of polymer lithium batteries for starting period and pulse power, J. Power Sources 54 (2) (1995) 393–396.
[22] S.C. Hageman, Simple PSpice Models Let You Simulate Common Battery Types, EDN, 1993, pp. 17–132.

[23] S. Gold, A PSpice macromodel for lithium-ion batteries, in: Proc. 12th Annu. Battery Conf. on Applications and Advances, 1997, pp. 215–222.
[24] S. Abu-Sharkh, D. Doerffel, Rapid test and non-linear model characterization of solid-state lithium-ion batteries, J. Power Sources 130 (2004) 266–274.
[25] J. Manwell, J. McGowan, Lead acid battery storage model for hybrid energy systems, Sol. Energy 50 (1993) 399–405.
[26] S.C. Hageman, Simple PSpice models let you simulate common battery types, Electronic Design News 38 (1993) 117–129.
[27] K. Siczek, Possible Use of the Bacterial Battery for Electric Motors (SAE Technical Paper 2014-01-2896), 2014, https://doi.org/10.4271/2014-01-2896.
[28] M. Doyle, T.F. Fuller, J. Newman, Modeling of galvanostatic charge and discharge of the lithium/polymer/insertion cell, J. Electrochem. Soc. 140 (6) (1993) 1526–1533.
[29] T.F. Fuller, M. Doyle, J. Newman, Simulation and optimization of the dual lithium ion insertion cell, J. Electrochem. Soc. 141 (1) (1994) 1–10.
[30] T.F. Fuller, M. Doyle, J. Newman, Relaxation phenomena in lithium-ion-insertion cells, J. Electrochem. Soc. 141 (4) (1994) 982–990.
[31] D. Rakhmatov, S. Vrudhula, An analytical high-level battery model for use in energy management of portable electronic systems, in: Proceedings of the International Conference on Computer Aided Design (ICCAD'01), 2001, , pp. 488–493.
[32] C. Chiasserini, R. Rao, Pulsed battery discharge in communication devices, in: Proceedings of the 5th International Conference on Mobile Computing and Networking, 1999, , pp. 88–95.
[33] C. Chiasserini, R. Rao, A model for battery pulsed discharge with recovery effect, in: Wireless Communications and Networking Conference, 1999, , pp. 636–639.
[34] C. Chiasserini, R. Rao, Improving battery performance by using traffic shaping techniques, IEEE J. Sel. Areas Commun. 19 (7) (2001) 1385–1394.
[35] H.-G. Schweiger, O. Obeidi, o. Komesker, A. Raschke, M. Schiemann, C. Zehner, M.Gehnen, M. Keller, P. Birke, Comparison of several methods for determining the internal resistance of lithium ion cells, Sensors (Basel) 10 (6) (2010) 5604–5625.
[36] A.A. Pesaran, D.J. Russel, J.W. Crawford, R. Rehn, E.A. Lewis, A Unique Calorimeter-Cycler for Evaluating High-Power Battery Modules, National Renewable Energy Laboratory, Golden, CO, 1997.
[37] A.A. Pesaran, Battery thermal management in EVs and HEVs: issues and solutions, in: Proceedings of the Advanced Automotive Battery Conference, Las Vegas, NV, 6–8 February, 2001.
[38] B.V. Ratnakumar, M.C. Smart, M.C. Whitcanack, M.C. Ewell, The impedance characteristics of Mars exploration rover Li-ion batteries, J. Power Sources 159 (2006) 1428–1439.
[39] J. Broadhead, H.C. Kuo, Electrochemical principals and reactions, in: D. Linden, T.B.Reddy (Eds.), Handbook of Batteries, third ed., McGraw-Hill, New York, NY, 1999.
[40] W. Weydanz, A. Jossen, Moderne Akkumulatoren richtig einsetzen (Print Your Book), Darmstadt, Germany, 2006.
[41] D. Berndt, Bleiakkumulatoren, 11th ed., VDI Verlag GmbH, Düsseldorf, Germany, 1986.
[42] F. Huet, A review of impedance measurements for determination of the state-of-charge or state-of-health of secondary batteries, J. Power Sources 70 (1998) 59–69.
[43] J.B. Copetti, F. Chenlo, A review of impedance measurements of the state-of-charge or state-of-health of secondary batteries, J. Power Sources 47 (1994) 109–118.
[44] BS 6290-4:1997, Lead-Acid Stationary Cells and Batteries. Specification for Classifying Valve Regulated Types, British Standards Institution, London, UK, 1997.
[45] Stationary Lead-Acid Batteries—Part 11, Vented Types—General Requirements and Methods of Test, IEC, Geneva, Switzerland, 2002.

[46] Y. Kobayashi, H. Miyashiro, K. Kumai, K. Takei, T. Iwahori, I. Uchidab, Precise electrochemical calorimetry of $LiCoO_2$/graphite lithium-ion cell, understanding thermal behavior and estimation of degradation mechanism, J. Electrochem. Soc. 149 (2002) A978–A982.
[47] V.S. Donepudi, B.E. Conway, Electrochemical calorimetry of the zinc and bromine electrodes in zinc-bromine and zinc-air batteries, J. Electrochem. Soc. 131 (1984) 1477–1485.
[48] E. Karden, Using Low-Frequency Impedance Spectroscopy for Characterization, Monitoring and Modelling of Industrial Batteries, Shaker Verlag GmbH, Aachen, Germany, 2001.
[49] V.G. Kumar, N. Munichandraiah, A.K. Shukla, Electrode impedance parameters and internal resistance of a sealed nickel/metal-hydride cell, J. Power Sources 63 (1996) 203–208.
[50] V. Muenzel, A.F. Hollenkamp, A.I. Bhatt, J. de Hoog, M. Brazil, D.A. Thomas, I. Mareels, A comparative testing study of commercial 18650-format lithium-ion battery cells, J. Electrochem. Soc. 162 (8) (2015) A1592–A1600, https://doi.org/10.1149/2.0721508jes.
[51] T. Zhang, M. Marinescu, L. O'Neill, M. Wild, G. Offer, Modeling the voltage loss mechanisms in lithium–sulfur cells: the importance of electrolyte resistance and precipitation kinetics, Phys. Chem. Chem. Phys. 17 (2015) 22581–22586.
[52] Y.V. Mikhaylik, J.R. Akridge, Polysulfide shuttle study in the Li/S battery system, J. Electrochem. Soc. 151 (2004) A1969–A1976.
[53] K. Kumaresan, Y.V. Mikhaylik, R.E. White, A mathematical model for a lithium–sulfur cell, J. Electrochem. Soc. 155 (2008) A576–A582.
[54] M. Ghaznavi, P. Chen, Sensitivity analysis of a mathematical model of lithium–sulfur cells: part I: applied discharge current and cathode conductivity, J. Power Sources 257 (2014) 394–401.
[55] M. Ghaznavi, P. Chen, Sensitivity analysis of a mathematical model of lithium–sulfur cells: part II: precipitation reaction kinetics and sulfur content, J. Power Sources 257 (2014) 402–411.
[56] M. Ghaznavi, P. Chen, Analysis of a mathematical model of lithium-sulfur cells: part III: electrochemical reaction kinetics, transport properties and charging, Electrochim. Acta 137 (2014) 575–585.
[57] D.N. Fronczek, W.G. Bessler, Insight into lithium–sulfur batteries: elementary kinetic modeling and impedance simulation, J. Power Sources 244 (2013) 183–188.
[58] A.F. Hofmann, D.N. Fronczek, W.G. Bessler, Mechanistic modeling of polysulfide shuttle and capacity loss in lithium–sulfur batteries, J. Power Sources 259 (2014) 300–310.
[59] V. Kolosnitsyn, E. Kuzmina, E. Karaseva, S. Mochalov, A study of the electrochemical processes in lithium–sulphur cells by impedance spectroscopy, J. Power Sources 196 (2011) 1478–1482.
[60] Z. Deng, Z. Zhang, Y. Lai, J. Liu, J. Li, Y. Liu, Electrochemical impedance spectroscopy study of a lithium/sulfur battery: modeling and analysis of capacity fading, J. Electrochem. Soc. 160 (2013) A553–A558.
[61] N.A. Canas, K. Hirose, B. Pascucci, N. Wagner, K.A. Friedrich, R. Hiesgen, Investigations of lithium–sulfur batteries using electrochemical impedance spectroscopy, Electrochim. Acta 97 (2013) 42–51.
[62] V.S. Kolosnitsyn, E.V. Kuz'mina, E.V. Karaseva, S.E. Mochalov, Impedance spectroscopy studies of changes in the properties of lithium-sulfur cells in the course of cycling, Russ. J. Electrochem. 47 (2011) 793–798.
[63] C. Barchasz, J.-C. Lepretre, F. Alloin, S. Patoux, New insights into the limiting parameters of the Li/S rechargeable cell, J. Power Sources 199 (2012) 322–330.

[64] D.R. Chang, S.H. Lee, S.W. Kim, H.T. Kim, Binary electrolyte based on tetra(ethylene glycol) dimethyl ether and 1,3-dioxolane for lithium–sulfur battery, J. Power Sources 112 (2002) 452–460.
[65] H. Yamin, E. Peled, Electrochemistry of a nonaqueous lithium/sulfur cell, J. Power Sources 9 (1983) 281–287.
[66] J.P. Neidhardt, D.N. Fronczek, T. Jahnke, T. Danner, B. Horstmann, W.G. Bessler, A flexible framework for modeling multiple solid, liquid and gaseous phases in batteries and fuel cells, J. Electrochem. Soc. 159 (2012) A1528–A1542.
[67] M. Marinescu, T. Zhang, G.J. Offer, A zero dimensional model of lithium–sulfur batteries during charge and discharge, Phys. Chem. Chem. Phys. 18 (2016) 584–593.
[68] A. Fotouhi, D.J. Auger, K. Propp, S. Longo, Electric vehicle battery parameter identification and SOC observability analysis: NiMH and Li-S case studies, IET Power Electron. (2017).
[69] A. Fotouhi, D.J. Auger, K. Propp, S. Longo, R. Purkayastha, L. O'Neill, S. Walus, Lithium-sulfur cell equivalent circuit network model parameterization and sensitivity analysis, IEEE Trans. Veh. Technol. 66 (2017) 7711–7721.
[70] V. Knap, D. Stroe, M. Swierczynski, R. Teodorescu, E. Schaltz, Investigation of the self-discharge behavior of lithium-sulfur batteries, J. Electrochem. Soc. 163 (2016) A911–A916.
[71] K. Propp, M. Marinescu, D. Auger, L. O'Neill, A. Fotouhi, K. Somasundaram, G. Offer, G. Minton, S. Longo, M. Wild, et al., Multi-temperature state-dependent equivalent circuit discharge model for lithium-sulfur batteries, J. Power Sources 328 (2016) 289–299.
[72] V. Knap, D.-I. Stroe, M. Swierczynski, R. Purkayastha, K. Propp, R. Teodorescu, E. Schaltz, A self-discharge model of lithium-sulfur batteries based on direct shuttle current measurement, J. Power Sources 336 (30) (2016) 325–331.
[73] G.L. Plett, Extended Kalman filtering for battery management systems of LiPB-based HEV battery packs—part 1. Background, J. Power Sources 134 (2004) 252–261.
[74] G.L. Plett, Extended Kalman filtering for battery management systems of LiPB-based HEV battery packs—part 2. Modelling and identification, J. Power Sources 134 (2004) 262–276.
[75] G.L. Plett, Extended Kalman filtering for battery management systems of LiPB-based HEV battery packs—part 3. State and parameter estimation, J. Power Sources 134 (2004) 277–292.
[76] K. Propp, D.J. Auger, A. Fotouhi, S. Longo, V. Knap, Kalman-variant estimators for state of charge in lithium-sulfur batteries, J. Power Sources 343 (2017) 254–267.
[77] A. Fotouhi, D.J. Auger, K. Propp, S. Longo, Lithium-sulfur battery state-of-charge observability analysis and estimation, IEEE Trans. Power Electron. (2017).
[78] A. Fotouhi, D.J. Auger, L. O'Neill, T. Cleaver, S. Walus, Lithium-sulfur battery technology readiness and applications—a review, Energies 10 (2017) 1937, https://doi.org/10.3390/en10121937.
[79] F.H. Zahrul, W.C. Lee, F.M.S. Mohd, B.I. Amir, Modeling of sodium sulfur battery for power system applications, Elektrika 9 (2) (2007) 66–72.
[80] H.L. Chan, D. Sutanto, A new battery model for use with battery energy storage system and electrical vehicles power systems, in: Proc. IEEE PES Winter Meeting, January 23–27, vol. 1, 2000, pp. 470–475.
[81] Y.-H. Kim, H.-D. Ha, Design of interface circuits with electrical battery models, IEEE Trans. Ind. Electron. 44 (1) (1997).

CHAPTER SEVENTEEN

Methods and Equipment for Measurement of Battery Parameters

According to Ref. [1], there is no commercially available battery management system for Li-S batteries, and there are no published methods for determining state of charge (SoC) in situ. The applicability of "standard" lithium-ion SoC estimation methods for Li-S batteries was explored. Open-circuit voltage methods and Coulomb counting were found to have a poor fit for lithium-sulfur, and model-based methods, particularly recursive Bayesian filters, were identified as showing strong promise. Three recursive Bayesian filters were implemented: an extended Kalman filter (EKF), an unscented Kalman filter (UKF) and a particle filter (PF). These estimators were tested through practical experimentation, considering both a pulse-discharge test and a test based on the New European Driving Cycle (NEDC). The experiment was carried out at a constant temperature, mirroring the environment expected in the authors' target automotive application. Estimators based on a relatively simple equivalent-circuit-network model delivered useful results. Of the three estimators implemented, the unscented Kalman filter gave the most robust and accurate performance, with an acceptable computational effort.

In Ref. [2] a nonlinear SoC dependent Li-S equivalent circuit network (ECN) model for a Li-S cell under discharge was presented. Li-S batteries are different from Li-ion batteries and require chemistry-specific models. A Li-S model was obtained using a "behavioral" interpretation of the ECN model; because Li-S exhibited a steep open-circuit voltage (OCV) profile at high SoCs, identification methods are designed to consider OCV changes during current pulses. The prediction-error minimization technique was used. The model was parameterized from experiments using a mixed-size current pulse profile at four temperatures from 10°C to 50°C, giving linearized ECN parameters for a range of SoCs, currents, and temperatures. They were used to create a nonlinear polynomial-based battery model suitable for use in a battery management system. Using the mentioned model to predict the

Next-generation Batteries with Sulfur Cathodes
https://doi.org/10.1016/B978-0-12-816392-4.00017-7

behavior of a validation data set representing an automotive NEDC driving cycle, the accurate terminal voltage predictions with a root mean square (RMS) error of 32 mV were obtained.

According to Ref. [3], rate-capability tests are often used for Li-S systems. It was found, from study on the individual effects of several cycling- and electrolyte-related parameters on a simple Li-S cell, that the rate-capability results are sensitive to the applied current and the cycling prehistory of the cell. The performance depends on the order in which cycling rate is changed. Slow initial rates aggravate material loss and reduce the achievable capacity. The results of rate-capability tests are different when the rates are varied only on the charge or only on the discharge. The charge rate does not directly affect the measured capacity, whereas the discharge rate does, especially when discharge cutoff voltages are high, due to slow discharge reactions. High charge rates affect the long-term stability, which is difficult to predict from usual rate-capability tests.

The study described in Ref. [4] considered application-oriented models of Li-S cells. Existing ECN models often neglect self-discharge, but this can be important in applications. The self-discharge phenomenon is investigated for a new 21 Ah Li-S cell using an ECN model, which was extended to account for cells' self-discharge. Formal system identification techniques were used to parameterize a model from experimental data. The original model was then extended by adding terms to represent a self-discharge resistance. To obtain the self-discharge resistance, a new series of experiments were designed and performed on the Li-S cell at various temperature and initial SoC levels.

Battery testing is possible using different current profiles. As a case study, a current profile was used based on EV power demand on urban dynamometer driving schedule (UDDS), also known as U.S. FTP-72 (Federal Test Procedure) [5]. The main feature of such a profile is the random change of current value in a realistic scenario. To obtain this current profile, a typical EV (i.e., Nissan Leaf) was simulated on a UDDS drive cycle, as discussed in Ref. [6]. The power demand signal was then scaled-down to be applied to the Li-S cell. Since this is a prototype cell, the current was limited below 10 A. This did not affect the results because the final cells had the same characteristic curve shapes. The scaled current corresponding to the UDDS velocity profile was applied to a fully charged cell and this was repeated until the cell became depleted. The temperature was controlled at 20°C during this test.

According to Ref. [7], most of the recent studies regarding the self-discharge behavior of Li-S batteries were focused only on simple comparisons between well-established and newly developed coin cells at one or two conditions (e.g., temperature value, depth-of-discharge, etc.) [8–10]. In a similar manner, a study on a variety of sulfur electrode materials was conducted in Ref. [11], where the reversible and irreversible capacity losses of the materials were identified. The self-discharge characteristics of Li-S coin cells were extensively studied through open circuit voltage (OCV) measurement, electrochemical impedance spectroscopy (EIS), and discharge curve at 25°C in Refs. [12,13].

During studies described in Ref. [12], on the self-discharge behaviors of the Li-S battery (Li/TEGDME/S), the changes of OCV and discharge curves were measured as a function of storage time. The battery was discharged to 1.7 V at a constant specific current of $100\,mA\,g^{-1}$ (based on sulfur) at 25°C. The AC impedance measurements were performed with amplitude of 10 mV over a frequency range of 10^{5}–10^{-2} Hz using a CMS100 electrochemical measurement system (Gamry Instruments Inc.).

During studies on Li-S cells described in Ref. [13], variations in the open circuit voltage (OCV) and electrochemical impedance curves of the studied Li-S cell with a pure sulfur cathode were measured as a function of storage time. The AC impedance measurements were performed by a frequency response analyzer (FRA) technique on an Autolab Electrochemical Workstation (PGSTAT30, Utrecht, Netherlands) over the frequency range from 100 kHz to 10 mHz with an amplitude of ±10 mV. The first discharge capacity of the cells was measured before and after 30 days of storage at 25°C. The cells were discharged to 1.5 V (vs. Li/Li^{+}) at a constant current density of $100\,mA\,g^{-1}$ (based on sulfur) with a battery test system (BTS, Tehran, Iran). All the specific capacity values were calculated based on sulfur mass. After the discharge tests were completed, the aged cells were disassembled, and the surface of the aluminum current collectors was observed using an optical microscope to determine the intensity of corrosion.

During studies on a prismatic Li-S cell described in Ref. [14], electrical testing was performed with a Digatron multiple battery tester MBT 01-05-16.

During studies on the newer generation of Sion Power Li-S cells described in Ref. [15], the dependence of the self-discharge on the storage time (in the range of days and months) was examined only for two depth-of-discharge (DOD) levels, i.e., 0% and 60%, at 20°C.

In the study described in Ref. [7], the capacity loss during storage was separated into the irreversible and reversible capacity loss. The study considered the influence of the DOD, idling time, and temperature on the self-discharge characteristics of the studied 3.4 Ah Li-S pouch cells. The investigation used open circuit voltage (OCV) measurements and discharge voltage curves for determining the self-discharge characteristic of the considered Li-S battery cells. Based on the experimental results, an estimation of the remaining battery cell capacity was performed.

All tests were performed using Digatron BTS 600 and MACCOR Series 4000 test stations.

During all the tests, the cells were placed inside temperature chambers with controlled environment temperature at 15, 25, 35, and 45°C.

The standard test protocol, which was used for the measurement of the self-discharge of the considered Li-S cells, is shown in Fig. 64.

The test protocol was composed of three steps, as follows:

- *Step 1*—precondition cycle on a fully discharged cell (charging: current of 0.1 C-rate (0.34 A), cut-off voltage 2.45 V or 11 h; discharging: 0.2 C-rate (0.68 A), cut-off voltage 1.5 V) to have the cell in a comparable state between the tests and to obtain the actual discharge cell capacity (C_{ini});

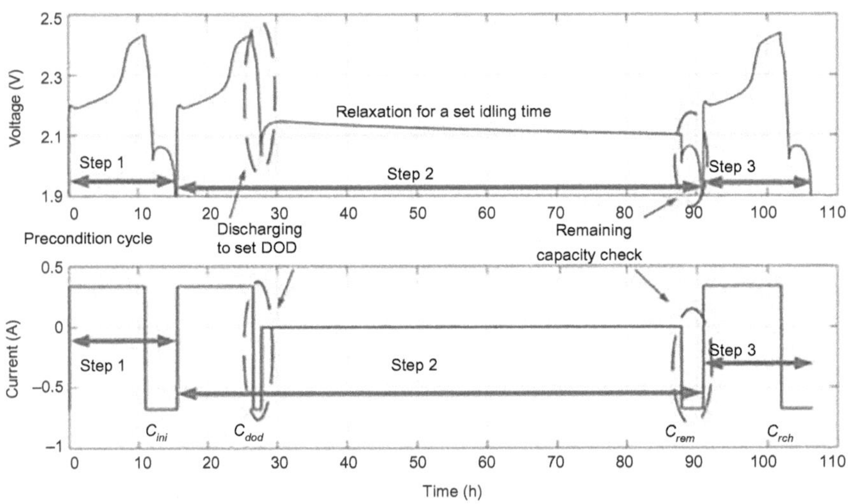

Fig. 64 Standard test protocol for systematic self-discharge measurement. *(Taken from V. Knap, D.L. Stroe, M.J. Swierczynski, R. Teodorescu, E. Schaltz, Investigation of the self-discharge behavior of lithium-sulfur batteries, Electrochem. Soc. J. 163(6) (2016) A911–A916, doi:10.1149/2.0641606jes.)*

- *Step 2*—the cell was fully charged and later discharged by the corresponding capacity (C_{dod}) to a predefined DOD value, where the cell was kept at open-circuit conditions for a certain idling time. Finally, after this idling time, the battery cell was discharged to measure the remaining cell capacity (C_{rem}).
- *Step 3*—the cell was recharged in a similar way to Step 1, to identify the new actual discharge capacity of the cell (C_{rch}), which allows the irreversible capacity lost due to calendar and cycling aging to be identified.

This self-discharge test procedure was repeated for the considered DOD levels, temperature levels, and idling times.

The self-discharge behavior was quantified based on the methodology presented in Ref. [15], which allows the reversible and irreversible capacity loss to be separated, as illustrated in Fig. 65 and computed using Eqs. (29)–(32):

$$C_t = (C_{ini} - C_{dod} - C_{rem})/C_{ini}{}^{*}100 \tag{29}$$

$$C_{sd} = (C_{rch} - C_{rem} - C_{dod}))/C_{ini}{}^{*}\,100 \tag{30}$$

$$C_{ir} = (C_{ini} - C_{rch})/C_{ini}{}^{*}100 \tag{31}$$

$$C_t = C_{sd} + C_{ir} \tag{32}$$

where:

C_t—the total capacity loss during the idling,

C_{ini}—the initial discharge capacity,

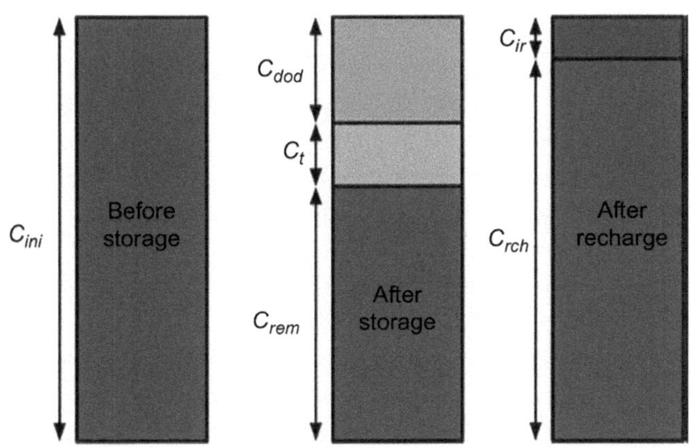

Fig. 65 Illustration of self-discharge quantification and separation. *(Taken from V. Knap, D.L. Stroe, M.J. Swierczynski, R. Teodorescu, E. Schaltz, Investigation of the self-discharge behavior of lithium-sulfur batteries, Electrochem. Soc. J. 163(6) (2016) A911–A916, doi:10.1149/2.0641606jes.)*

C_{dod}—the discharged capacity to the specific DOD point,
C_{rem}—the remaining capacity after the idling time,
C_{sd}—reversible capacity loss, which is further referred to as the self-discharge rate,
C_{rch}—the new actual discharge capacity during recharge after the idling,
C_{ir}—the irreversible capacity loss.

The irreversible capacity loss was caused by cycling and idling (storage) degradation of the Li-S cells.

According to Ref. [14], the actual capacity of the high voltage plateau (C_H), (Fig. 66) can be expressed by Eq. (33).

$$C_H = C_{H\_ini}{}^{*} \exp[-(k_S/t_S)] \quad (33)$$

where:

C_{H_ini}—the initial discharge capacity corresponding to the high plateau,
k_S—the self-discharge constant,
t_S—the idling time.

The self-discharge constant k_S can be determined experimentally, as the slope of the line describing the variation of $\ln(C_H/C_{H\_ini})$ with the idling time t_S.

According to Ref. [16], degradation of lithium-ion batteries is caused by aging mechanisms grouped into three degradation modes (DMs): conductivity loss (CL), loss of active material (LAM), and loss of lithium inventory (LLI). State of Health (SoH) is typically the parameter used by the Battery Management System (BMS) to quantify battery degradation based on the decrease in capacity and the increase in resistance. The definition of SoH within a BMS does not include an indication of the underlying DMs causing the degradation. Some studies have analyzed the effects of the DMs using incremental capacity and differential voltage (IC-DV) and electrochemical impedance spectroscopy (EIS). After comparison of IC-DV and EIS on the

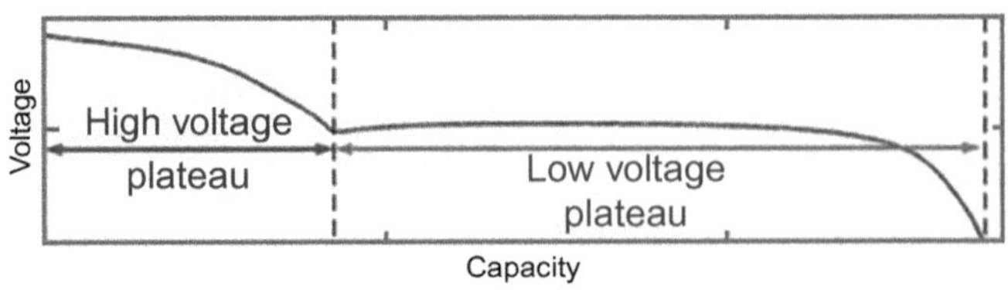

Fig. 66 Typical voltage discharging profile of a Li-S battery with marked high and low voltage plateaus. *(Taken from Y.V. Mikhaylik, J.R. Akridge, Polysulfide shuttle study in the LiÕS battery system, J. Electrochem. Soc. 151 (2004) A1969.)*

same data, the effects due to LAM and LLI were found to be the most pertinent, outlining that both techniques are correlated.

According to Ref. [17], the electrochemical behavior of Li-S batteries was studied by means of EIS. Measurements were performed in equidistant charge intervals at different depths of discharge and charge during the first cycle. The degradation of the cells was analyzed for up to 50 cycles. An equivalent electrical circuit was used to simulate the electrochemical processes and to quantify the impedance contributions of Li-S batteries. Atomic force microscopy (AFM) was used to provide information about changes in the electrical conductivity of the cathode surface related to the building of isolating films.

According to Ref. [18], the internal resistance is usually calculated by the EIS method, giving unrealistic low internal resistance values. In the paper internal resistance was calculated from the voltage drop with the FreedomCAR method, where the cell was represented by an ideal voltage source with two internal resistances and two capacitors. The validation of the results from the FreedomCAR method was much better (99%) than the EIS method. The behavior of individual cells was examined while they worked together in a module.

During studies described in Ref. [19], the EIS and cyclic voltammograms (CVs) of the cells were used, with coaxial graphene-coated cotton-carbon (CGCC) cathodes to understand the relationship between the CGCC configuration and the cell characteristics.

During studies performed in Ref. [20], galvanostatic charge-discharge measurements were carried out in a potential range between 1.7 and 2.5 V vs. Li/Li^+ using a MACCOR (Tulsa, Oklahoma) battery cycler. After 1 cycle at C/50 (with $1C = 1672\,mA\,g(S)^{-1}$) was completed, the cells were cycled at charge and discharge rates of C/8 and C/5, respectively. At such C-rates, high specific capacities can be achieved while keeping the cell failure rate at a minimum. All electrochemical experiments were conducted under stable environmental conditions in a BINDER cooled incubator. The electrolyte-to-sulfur mass ratios were 20:1 and 10:1 for coin-type and pouch cells, respectively.

Additionally, field-emission scanning electron microscopy (SEM) images were recorded with a LEO 1530 instrument at 10 kV. In operando X-ray diffraction (XRD) measurements were carried out at the Synchrotron Light Source ANKA on the PDIFF beamline (wavelength of 0.08856 nm, beam size of 1.5 mm (vertical) × 0.25 mm (horizontal), sample-to-detector distance of 215.402 mm) using a Pilatus 300k detector (counting time of 30 s).

Tests of Li-S cells described in Ref. [21] covered various battery operating conditions. All the tests were carried out on 3.4 Ah long-life Li-S cells produced by OXIS Energy Ltd. The Series-4000 battery tester, a voltage/current device that applies a current profile and measures the voltage, or vice versa, was used for the experiments. Li-S cells were contained inside an aluminum test box connected to the equipment using crocodile clips. The boxes were inside a Binder thermal chamber to set the desired temperature during each test. A mixed charge-discharge pulse test (Fig. 67) was used for Li-S cell model parameterization. Pulse tests are commonly used for modeling of various battery types. The pulse tests used in the study were based on those used by the cell manufacturer.

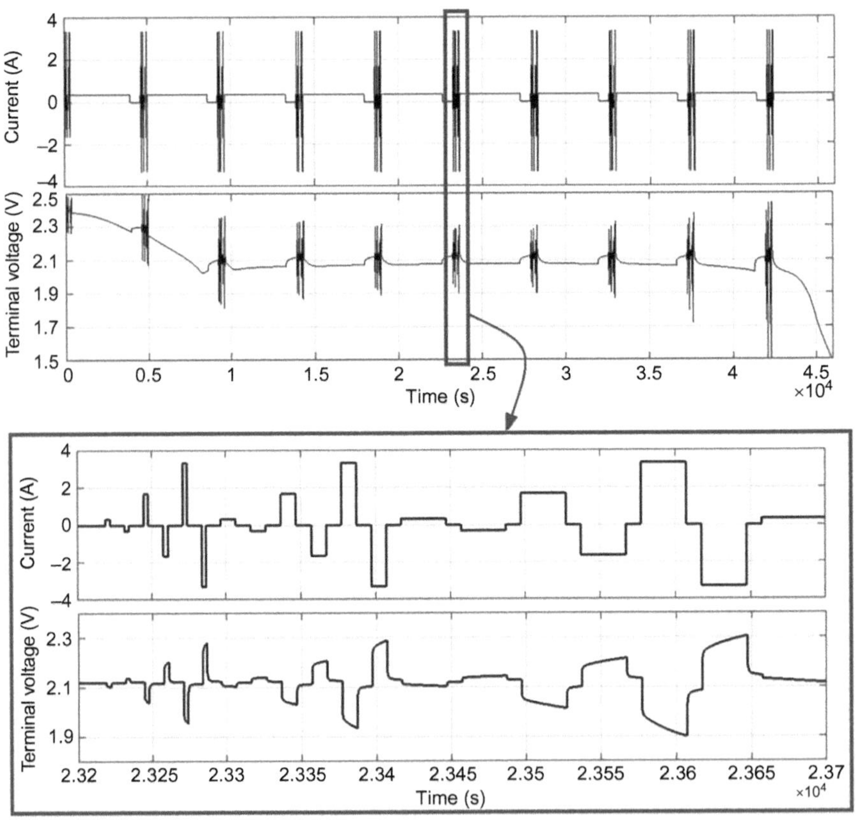

Fig. 67 Mixed charge/discharge impulse test. *(Taken from A. Fotouhi, D.J. Auger, K. Propp, S. Longo, R. Purkayastha, L. O'Neill, S. Walus, Lithium–sulfur cell equivalent circuit network model parameterization and sensitivity analysis, IEEE Trans. Veh. Technol. 66(9) (2017).)*

In the test, various charge/discharge rates and different SoC levels were considered. The test started at full charge state (2.45 V) and continued until the cell's terminal voltage dropped below 1.5 V (i.e., the cut-off voltage), which means a depleted state. Consecutive charge/discharge current pulses were applied to the cell, as shown in Fig. 6. The pulse sequence consists of 18 pulses (9 discharge and 9 charge pulses) including different frequencies and amplitudes. The whole pulse sequence was applied ten times at ten charge levels to investigate the effects of SoC. The models derived in the study represent a mix of discharge and charging behavior: discharge is dominant, but there is a small amount of charging. This is chosen to simulate an EV driving scenario in which regenerative braking is modeled with charge pulses. Hysteresis effects are not explicitly modeled. The maximum current is below 1C. Data was collected in the time domain with a sampling rate of 1 s. The measurements included time, current, and the cell's terminal voltage while temperature was monitored to ensure that it is being kept constant by the test equipment. The test was repeated at various temperature levels including 10, 20, 30, 40 and 50°C. Model parameterization was performed in each case.

According to Ref. [22], during the dissolution-precipitation process in a Li-S cell, the total volume change of the electrolyte in the pore space occurs due to precipitation/dissolution of the solid sulfur phase and due to the cathode microstructure shrinking or swelling to accommodate the changes in the pore volume resulting from the electrolyte induced hydrostatic pressure. Current Li-S performance models neglect this contribution. A computational methodology was developed to quantify the impact of precipitation-induced volume change, pore morphology, and confinement attributes in a Li-S cathode.

To study morphological aspects of Li-S cells, X-ray tomography can be useful.

According to Ref. [23], the study of morphological changes is crucial to understand degradation phenomena in Li-S batteries.

As reported in Ref. [24], X-ray tomography was applied to investigate individual parts of their Li-S battery in terms of porosity, surface structure and confined sulfur after 50 cycles.

During studies described in Ref. [25], synchrotron two-dimensional X-ray transmission microscopy was used to investigate their sulfur battery electrodes, in operando, during different stages of one cycle.

According to Ref. [23], in contrast to conventional lab tomography systems, the highly coherent monochromatic synchrotron X-rays in combination with the phase contrast technique are suited to study actual particle sizes, contact areas, and small features, like cracks in the sulfur particles.

X-ray tomography was used for the exploration of the microstructure and degradation of other types of batteries such as alkaline batteries or lithium-based electrodes [26–29].

In Ref. [23] three-dimensional synchrotron X-ray phase contrast tomography was applied to extract morphological parameters influencing degradation of a Li-S cell. In contrast to two-dimensional imaging techniques, this method allows large, representative volumes to be accessed. In these volumes three-dimensional parameters like contact areas between the phases, size distributions, and spatial distributions throughout the whole electrode height can be quantified in dependence on the aging state of the battery electrode.

In Ref. [30] a critical analysis was given of the various AC "impedance" and DC "resistance" or "voltage response" measurement methods used in both portable (off-line) battery testing equipment and stationary (on-line) battery monitoring systems. A close-to-ideal method based on pulsed DC impedance was presented therein.

According to Ref. [18], batteries are often tested per cell. But in most cases more than one single cell is needed for an application and the characteristics of a module of cells is not the same. In other cases, the whole module is examined as one big cell, without looking at the individual cells. But the weakest cell affects the performance of the whole module.

During studies described in Ref. [31], the room-temperature cycling characteristics of the Na-S cells were evaluated under galvanostatic conditions using Neware CT-3008 battery testers and electrochemical processes in the cells were studied by cyclic voltammetry using a CHI600D potentiostat. Electrochemical impedance and floating tests were conducted by using a Solartron Cell Test System model 1470E potentiostat/galvanostat. Ionic conductivities were measured using a Novocontrol N40 broadband dielectric spectrometer.

REFERENCES

[1] K. Propp, D.J. Auger, A. Fotouhi, S. Longo, V. Knap, Kalman-variant estimators for state of charge in lithium-sulfur batteries, J. Power Sources 343 (1) (2017) 254–267.

[2] K. Propp, M. Marinescu, D.J. Auger, L. O'Neill, A. Fotouhi, K. Somasundaram, G. J. Offer, G. Minton, S. Longo, M. Wild, V. Knap, Multi-temperature state-dependent equivalent circuit discharge model for lithium-sulfur batteries, J. Power Sources 328 (1) (2016) 289–299.

[3] T. Poux, P. Novák, S. Trabesinger, Pitfalls in Li–S rate-capability evaluation, J. Electrochem. Soc. 163 (7) (2016) A1139–A1145, https://doi.org/10.1149/2.0181607jes.
[4] S.E.A. Yousif, A. Fotouhi, D.J. Augerz, K. Propp, Self-discharge effects in lithium-sulfur equivalent circuit networks for state estimation, J. Electrochem. Soc. 165 (1) (2018) A6081–A6090, https://doi.org/10.1149/2.0121801jes.
[5] Anon, Emissions Cycles: FTP-72 (UDDS), DieselNet (Website), https://www.dieselnet.com/standards/cycles/ftp72.php (Accessed August 21, 2017).
[6] A. Fotouhi, N. Shateri, D.J. Auger, S. Longo, K. Propp, R. Purkayastha, M. Wild, MATLAB graphical user interface for battery design and simulation; from cell test data to real-world automotive simulation, in: SMACD2016, Lisbon, 2016.
[7] V. Knap, D.L. Stroe, M.J. Swierczynski, R. Teodorescu, E. Schaltz, Investigation of the self-discharge behavior of lithium-sulfur batteries, Electrochem. Soc. J. 163 (6) (2016) A911–A916, https://doi.org/10.1149/2.0641606jes.
[8] N. Azimi, Z. Xue, N.D. Rago, C. Takoudis, M.L. Gordin, J. Song, D. Wang, Z. Zhang, Fluorinated electrolytes for Li-S battery: suppressing the self-discharge with an electrolyte containing fluoroether solvent, J. Electrochem. Soc. 162 (2015) 64–68.
[9] M.L. Gordin, F. Dai, S. Chen, T. Xu, J. Song, D. Tang, N. Azimi, Z. Zhang, D. Wang, Bis(2,2,2-trifluoroethyl) ether as an electrolyte co-solvent for mitigating self-discharge in lithium–sulfur batteries, ACS Appl. Mater. Interfaces 6 (2014) 8006–8010.
[10] J.-Q. Huang, T.-Z. Zhuang, Q. Zhang, H.-J. Peng, C.-M. Chen, F. Wei, Permselective graphene oxide membrane for highly stable and anti-self-discharge lithium–sulfur batteries, ACS Nano 9 (2015) 3002–3011.
[11] C.J. Hart, M. Cuisinier, X. Liang, D. Kundu, A. Garsuch, L.F. Nazar, Rational design of sulphur host materials for Li–S batteries: correlating lithium polysulphide adsorptivity and self-discharge capacity loss, Chem. Commun. 51 (2015) 2308–2311.
[12] H.S. Ryu, H.J. Ahn, K.W. Kim, J.H. Ahn, K.K. Cho, T.H. Nam, Self-discharge characteristics of lithium/sulfur batteries using TEGDME liquid electrolyte, Electrochim. Acta 52 (2006) 1563–1566.
[13] M. Kazazi, M.R. Vaezi, A. Kazemzadeh, Improving the self-discharge behavior of sulfur-polypyrrole cathode material by $LiNO_3$ electrolyte additive, Ionics 20 (2014) 1291, https://doi.org/10.1007/s11581-014-1095-2.
[14] Y.V. Mikhaylik, J.R. Akridge, Polysulfide shuttle study in the LiÕS battery system, J. Electrochem. Soc. 151 (2004)A1969.
[15] C.E. Parfitt, Characterisation, Modelling and Management of Lithium-Sulphur Batteries for Spacecraft Applications (Ph.D. Thesis), University of Warwick, 2012.
[16] C. Pastor-Fernández, K. Uddin, G.H. Chouchelamane, W.D. Widanage, J. Marco, A comparison between electrochemical impedance spectroscopy and incremental capacity-differential voltage as Li-ion diagnostic techniques to identify and quantify the effects of degradation modes within battery management systems, J. Power Sources 360 (2017) 301–318.
[17] N.A. Cañas, K. Hirose, B. Pascucci, N. Wagner, K.A. Friedrich, R. Hiesgen, Investigations of lithium–sulfur batteries using electrochemical impedance spectroscopy, Electrochim. Acta 97 (1) (2013) 42–51.
[18] H. Culcu, B. Verbrugge, N. Omar, P. Van Den Bossche, J. Van Mierlo, Internal resistance of cells of lithium battery modules with FreedomCAR model, in: EVS24 Stavanger, Norway, May 13–16, World Electric Vehicle Journal, AVERE, vol. 3, 2009, ISSN 2032-6653-2009.
[19] S.-H. Chung, A. Manthiram, Designing lithium-sulfur cells with practically necessary parameters, Joule 2 (2018).
[20] J. Kulisch, H. Sommer, T. Brezesinski, J. Janek, Simple cathode design for Li–S batteries: cell performance and mechanistic insights by in operando X-ray diffraction, Phys. Chem. Chem. Phys. 16 (2014) 18765–18771, https://doi.org/10.1039/C4CP02220C.

[21] A. Fotouhi, D.J. Auger, K. Propp, S. Longo, R. Purkayastha, L. O'Neill, S. Walus, Lithium–sulfur cell equivalent circuit network model parameterization and sensitivity analysis, IEEE Trans. Veh. Technol. 66 (9) (2017).
[22] P. Barai, A. Mistry, P.P. Mukherjee, Poromechanical effect in the lithium–sulfur battery cathode, Extreme Mech. Lett. 9 (Part 3) (2016) 359–370.
[23] L. Zielke, C. Barchasz, S. Waluś, F. Alloin, J.-C. Leprêtre, A. Spettl, V. Schmidt, A. Hilger, I. Manke, J. Banhart, R. Zengerle, S. Thiele, Degradation of Li/S battery electrodes on 3D current collectors studied using X-ray phase contrast tomography, Sci. Rep. 5 (2015) 10921, https://doi.org/10.1038/srep10921.
[24] G. Zhou, et al., A graphene–pure-sulfur sandwich structure for ultrafast, long-life lithium–sulfur batteries, Adv. Mater. 26 (2014) 625–631.
[25] J. Nelson, et al., In operando X-ray diffraction and transmission X-ray microscopy of lithium sulfur batteries, J. Am. Chem. Soc. 134 (2012) 6337–6343.
[26] M. Ebner, F. Marone, M. Stampanoni, V. Wood, Visualization and quantification of electrochemical and mechanical degradation in Li ion batteries, Science 342 (2013) 716–720.
[27] P. Shearing, R. Bradley, J. Gelb, N. Brandon, P. Withers, Multi-length scale X-ray tomography of Li-ion batteries, Microsc. Microanal. 17 (2011) 1672–1673.
[28] I. Manke, et al., In situ investigation of the discharge of alkaline Zn–MnO_2 batteries with synchrotron X-ray and neutron tomographies, Appl. Phys. Lett. 90 (2007)214102.
[29] L. Zielke, et al., A combination of X-ray tomography and carbon binder modeling: reconstructing the three phases of $LiCoO_2$ Li-ion battery cathodes, Adv. Energy Mater. 4 (2014)1301617.
[30] J. O'Conner, T. Leonard, G. Bohensky, Comparing Methods to Determine the Health of Battery System, Thomas Leonard and Gene Bohensky Comp. Btech Inc., 2004.
[31] S. Wei, S. Xu, A. Agrawral, S. Choudhury, Y. Lu, Z. Tu, L. Ma, L.A. Archer, A stable room-temperature sodium–sulfur battery. Nat. Commun. 7 (2016)11722, https://doi.org/10.1038/ncomms11722.

CHAPTER EIGHTEEN

Recycling of Batteries With Sulfur Cathodes

Contents

Recycling of Li-S batteries is still insufficiently recognized. According to Ref. [1], to understand the environmental sustainability performance of a Li-S battery on future electric vehicles (Evs), a life cycle assessment (LCA) model was developed for comprehensive environmental impact assessment of a Li-S battery pack using a graphene sulfur composite cathode and a lithium metal anode protected by a lithium-ion conductive layer, for actual EV applications. The Li-S battery pack was configured with a 61.3 kWh capacity to power a midsize EV for a 320-km range. The life cycle inventory model was developed with a hybrid approach, based on lab-scale synthesis of the graphene sulfur composite. The impacts of the Li-S battery were assessed using the ReCiPe method and benchmarked with those of a conventional nickel-cobalt-manganese (NCM)-graphite battery pack under the same driving distance per charge. The environmental impact assessment results showed that the Li-S battery was more environmentally friendly than a conventional NCM-graphite battery, with 9%–90% lower impact. The improvement pathways for the Li-S battery to meet the USABC (U.S. Advanced Battery Consortium) targets were presented with the corresponding environmental impact changes.

As described in Ref. [2], in the case of lithium polysulfide batteries, the precipitation of insoluble and irreversible sulfide species on the surface of carbon and lithium (called "dead" sulfide species) leads to continuous capacity degradation in high mass loading cells, which represents a great challenge. A strategy was proposed to reactivate dead sulfide species by

Next-generation Batteries with Sulfur Cathodes
https://doi.org/10.1016/B978-0-12-816392-4.00018-9

reacting them with sulfur powder with stirring and heating (70°C) to recover the cell capacity, and a flow battery system based on the reactivation approach was further demonstrated. As a result, ultrahigh mass loading (0.125 g cm^{-3}, 2 g sulfur in a single cell), high volumetric energy density (135 Wh L^{-1}), good cycle life, and high single-cell capacity were achieved. The high volumetric energy density indicated its promising application for future grid energy storage.

An interesting proposition for use of a recycling method was presented in Ref. [3]. It was described that vulcanized rubber products contain polymer backbones crosslinked with sulfur to improve mechanical strength. Burning of waste rubber products emits toxic gases, and recycling of rubber by breaking the C—S bond with costly reagents and heat is also haunted by environmental concerns. The crosslinked polymers can be extracted chemically at room temperature and, by a simple solution-drop method, used to prepare a bifunctional cathode layer on the cathode current collector of a Li-S battery. The C—S bond of the crosslinking sulfur can be broken reversibly in a discharge/recharge cycle to provide a sulfur source, and the broken carbon bond can capture a lithium polysulfide soluble in a liquid organic electrolyte by forming a C—Li—S bond during discharge. During charge, the Li is extracted and the C—S bond is reformed. However, added sulfur powder is also needed in the cathode of a Li-S battery.

Also, in Ref. [4] a solution to produce high-performance Li-S batteries from recycled paper was reported. Common paper hardboard was peeled into long stripes and integrated with graphene oxide (GO) sheets by a capillary absorption method. The paper/GO hybrid was then activated at high temperature, forming activated paper carbon (APC)/graphene scaffold. The capillary absorbed GO sheets were homogeneously coated on the APC fibers and shrined geometrically during heating, forming nanosized bulges and pores, which remarkably improved sulfur loading and cathode conductivity. The recycled paper derived graphene Li-S battery exhibited superior life span of 620 cycles with an excellent capacity retention rate of 60.5%. More importantly, the APC/graphene/S stayed integral without pulverization even after such long cycling, indicating excellent mechanical robustness of the cathode material. Since the malfunction of the battery can be ascribed to the anode, an activated paper carbon (APC) interlayer was sandwiched between the Li anode and the separator to suppress the degradation of Li anode by protecting the Li from unfavorable reactions, stretching the battery lifespan up to 980 cycles with a capacitance retention rate of 52.3% and improved rate ability.

Some issues related to recycling of the Li-ion batteries can be like these of Li-S batteries, especially with respect to the anode materials. However, according to Ref. [5] at present, there are no regulations regarding recycling of large format Li-ion batteries.

Several recycling methods have been proposed, each with its advantages and disadvantages [6]. The three main types are as follows:

- pyrometallurgical recycling (smelting),
- intermediate recycling process,
- direct recycling.

The processes of Li-ion cell recycling must regain not only cobalt, nickel, copper, and aluminum from spent battery cells, but also a significant share of lithium. Therefore, some unit operations are combined to complex process chains [7,8].

Such unit operations are [7]:

Deactivation or discharging of the battery,

Disassembly of battery systems,

Mechanical processes (including crushing, sorting, and sieving processes [8]),

Hydrometallurgical processes,

Pyrometallurgical processes.

Specific dangers associated with lithium-ion battery recycling processes are: electrical dangers, chemical dangers, burning reactions, and their potential interactions [7]. A complicating factor is the water sensitivity.

Pouch cells are particularly easier to recycle in this way and sometimes this is done to salvage the copper despite the safety issues.

As of 2017, the recycling of Li-ion batteries generally does not extract lithium since the many different types of Li-ion batteries require different extraction processes [9]. Another reason why it isn't being done is because the extraction of lithium from old batteries is five times more expensive than mined lithium [10].

Perhaps they can be adapted to the case of Li-S cells, but further studies are needed.

Recycling of Na-S batteries is a separate issue.

According to Ref. [11], four potential treatment processes currently being evaluated by Sandia National Laboratories for Na/S battery disposal are:

- recovery of sodium polysulfides,
- oxidation to sodium sulfate, sodium chloride, and sulfur,
- recovery of sodium and sulfur,
- acidic oxidation and Claus process to form sulfur and sodium sulfate [12].

They are described further.

It was assumed that each Na/S battery would be disassembled into individual cells before processing. The battery would be discharged prior to cooling and disassembly, except for the process consisting of recovery of sodium and sulfur, in which case the battery would be fully charged before cooling and disassembly [12].

18.1 RECOVERY OF POLYSULFIDES

The sodium polysulfide is recovered by crushing the cells. Excess sulfur is added to oxidize any Na_2S_3 to Na_2S_4, followed by dissolving the sodium polysulfide in water. The aqueous solution is dried to a powder product and drummed for shipment. An alternative process would be to add excess sodium to the sodium polysulfide to reduce the Na_2S_3 to Na_2S.

18.2 RECOVERY OF SODIUM AND SULFUR

The battery is charged prior to disassembly of the cells. Although as many as 10% of the cells cannot be recharged, 90% of the cells will be charged. The individual cells are heated to 100°C and the sodium is removed. After removal of the sodium, the outer cell can is removed and the sulfur is hydrotreated to form H_2S. The hydrogen sulfide is oxidized to elemental sulfur via the Claus process.

18.3 OXIDATION TO FORM SODIUM SULFATE, SODIUM CHLORIDE, AND SULFUR

The battery is fully discharged to form Na_2S_3. (Battery discharge will not give 100% Na_2S_3 for all cells—some cells will be electrically isolated and this will have to be addressed.) The cell cases are breached and sodium hypochlorite (bleach) is added to oxidize the polysulfide to sodium sulfate and sulfur. After filtering, a solution of sodium sulfate and sodium chloride is produced.

18.4 ACIDIC OXIDATION AND CLAUS PROCESS TO FORM SULFUR AND SODIUM SULFATE

The battery is fully discharged to form Na_2S_3. The cell cases are breached, and sulfuric acid is added to the Na_2S_3 to form hydrogen sulfide and elemental sulfur. The hydrogen sulfide is oxidized to elemental sulfur via the Claus process.

REFERENCES

[1] Y. Deng, J. Li, T. Li, X. Gao, C.X. Yuan, Life cycle assessment of lithium sulfur battery for electric vehicles, J. Power Sources 343 (2017) 284–295.
[2] Y. Jin, G. Zhou, F. Shi, D. Zhuo, J. Zhao, K. Liu, y. Liu, C. Zu, W. Chen, R. Zhang, X. Huang, Y. Cui, Reactivation of dead sulfide species in lithium polysulfide flow battery for grid scale energy storage, Nat. Commun. 8 (2017)462 https://doi.org/10.1038/s41467-017-00537-0.
[3] B.-C. Yu, J.-W. Jung, K. Parka, J.B. Goodenough, A new approach for recycling waste rubber products in Li–S batteries, Energy Environ. Sci. 10 (2017) 86–90.
[4] Y. Zhang, Z. Gao, X. Li, Recycled Paper Derived Graphene Lithium-Sulfur Batteries with Enhanced Capacity and Extended Lifespan, Abstract MA2017-02 543.
[5] L. Gaines, The future of automotive lithium-ion battery recycling: charting a sustainable course, Sustain. Mater. Technol. 1–2 (2014) 2–7.
[6] J.B. Dunn, L. Gaines, M. Barnes, J. Sullivan, M. Wang, Material and Energy Flows in the Materials Production, Assembly, and End of Life Stages of the Automotive Lithium Ion Battery Life Cycle (Argonne National Laboratory Report ANL/ESD/12-3)2012,
[7] C. Hanisch, J. Diekmann, A. Stieger, W. Haselrieder, A. Kwade, J. Yan, L.F. Cabeza, R. Sioshansi (Eds.), Handbook of Clean Energy Systems—Recycling of Lithium-Ion Batteries, 5 Energy Storage ed., John Wiley & Sons, Ltd. ISBN 9781118991978, 2015, pp. 2865–2888 (Chapter 27).
[8] C. Hanisch, Recycling of Lithium-Ion Batteries, Presentation on Recycling of Lithium-Ion Batteries, Lion Engineering GmbH, 2015.
[9] Derfor er det vanskelig å resirkulere elbilbatterier, Teknisk Ukeblad, 2016.
[10] Kijk Magazine, 2017.
[11] D. Corbus, Environmental, health, and safety issues of sodium-sulfur for batteries electric and hybrid vehicles, in: Battery Recycling and Disposal, Prepared Under Task No. AS015440, vol. II, 1992.
[12] S.L. Lott, Presentation to Na/Beta Battery Reclamation Sub-Working Group, Washington, DC, May 29–30, 1991.

CHAPTER NINETEEN

Summary

Despite the development of lithium-ion batteries used widely in electric and hybrid cars and other applications, this technology has its drawbacks, which have not yet been eliminated or sufficiently limited. The search for new solutions to enable the most efficient collection of electricity is still underway. The various configurations of batteries using sulfur cathodes are very promising in this respect. Among these, the largest level of technical preparation is represented by Li-S batteries and they will soon be developed. Recently, many efforts have been made in various areas to modify or limit the impact of adverse physical phenomena (i.e., the shuttle effect of polysulfides) occurring in this type of battery. Na-S batteries are also a very promising direction of development, but they will be used rather as stationary energy storage. Other batteries using sulfur cathodes, like Mg-S, Zn-S, Al-S and others, need further study.

Next-generation Batteries with Sulfur Cathodes
https://doi.org/10.1016/B978-0-12-816392-4.00019-0

INDEX

Note: Page numbers followed by *f* indicate figures and *t* indicate tables.

C

M

P

T

MIX
Papier | Fördert gute Waldnutzung
FSC® C083411

Kolibri 360 GmbH
Ferdinand-Jühlke-Straße 7
99095 Erfurt, Deutschland
produktsicherheit@kolibri360.de